CLIMATE CHANGE EDUCATION

Formal Settings, K-14

A WORKSHOP SUMMARY

Alexandra Beatty, *Rapporteur*

Steering Committee on Climate Change Education
in Formal Settings, K-14

Board on Science Education

Division of Behavioral and Social Sciences and Education

NATIONAL RESEARCH COUNCIL
OF THE NATIONAL ACADEMIES

THE NATIONAL ACADEMIES PRESS
Washington, D.C.
www.nap.edu

THE NATIONAL ACADEMIES PRESS 500 Fifth Street, NW Washington, DC 20001

NOTICE: The project that is the subject of this report was approved by the Governing Board of the National Research Council, whose members are drawn from the councils of the National Academy of Sciences, the National Academy of Engineering, and the Institute of Medicine. The members of the committee responsible for the report were chosen for their special competences and with regard for appropriate balance.

This study was supported by Contract No. DUE-0956031 between the National Academy of Sciences and the National Science Foundation. Any opinions, findings, conclusions, or recommendations expressed in this publication are those of the author(s) and do not necessarily reflect the views of the organizations or agencies that provided support for the project.

International Standard Book Number-13: 978-0-309-26016-9
International Standard Book Number-10: 0-309-26016-7

Additional copies of this report are available from the National Academies Press, 500 Fifth Street, NW, Keck 360, Washington, DC 20001; (800) 624-6242 or (202) 334-3313; http://www.nap.edu.

Printed in the United States of America

Cover credits: The first and bottom photos are from iStock. The second photo from the top is of the Marian Koshland Science Museum of the National Academy of Sciences, Washington, DC. The third photo from the top is from the U.S. Fish and Wildlife Service website.

Suggested citation: National Research Council. (2012). *Climate Change Education in Formal Settings, K-14: A Workshop Summary*. A. Beatty, Rapporteur. Steering Committee on Climate Change Education in Formal Settings, K-14. Board on Science Education, Division of Behavioral and Social Sciences and Education. Washington, DC: The National Academies Press.

THE NATIONAL ACADEMIES

Advisers to the Nation on Science, Engineering, and Medicine

The **National Academy of Sciences** is a private, nonprofit, self-perpetuating society of distinguished scholars engaged in scientific and engineering research, dedicated to the furtherance of science and technology and to their use for the general welfare. Upon the authority of the charter granted to it by the Congress in 1863, the Academy has a mandate that requires it to advise the federal government on scientific and technical matters. Dr. Ralph J. Cicerone is president of the National Academy of Sciences.

The **National Academy of Engineering** was established in 1964, under the charter of the National Academy of Sciences, as a parallel organization of outstanding engineers. It is autonomous in its administration and in the selection of its members, sharing with the National Academy of Sciences the responsibility for advising the federal government. The National Academy of Engineering also sponsors engineering programs aimed at meeting national needs, encourages education and research, and recognizes the superior achievements of engineers. Dr. Charles M. Vest is president of the National Academy of Engineering.

The **Institute of Medicine** was established in 1970 by the National Academy of Sciences to secure the services of eminent members of appropriate professions in the examination of policy matters pertaining to the health of the public. The Institute acts under the responsibility given to the National Academy of Sciences by its congressional charter to be an adviser to the federal government and, upon its own initiative, to identify issues of medical care, research, and education. Dr. Harvey V. Fineberg is president of the Institute of Medicine.

The **National Research Council** was organized by the National Academy of Sciences in 1916 to associate the broad community of science and technology with the Academy's purposes of furthering knowledge and advising the federal government. Functioning in accordance with general policies determined by the Academy, the Council has become the principal operating agency of both the National Academy of Sciences and the National Academy of Engineering in providing services to the government, the public, and the scientific and engineering communities. The Council is administered jointly by both Academies and the Institute of Medicine. Dr. Ralph J. Cicerone and Dr. Charles M. Vest are chair and vice chair, respectively, of the National Research Council.

www.national-academies.org

STEERING COMMITTEE ON CLIMATE CHANGE EDUCATION IN FORMAL SETTINGS, K-14

CHARLES W. "ANDY" ANDERSON (*Chair*), Department of Teacher Education, Michigan State University
CAROL BREWER, Department of Biology, University of Montana
LYNN ELFNER, Ohio Academy of Science
JAMES E. GERINGER, Environmental System Research Institute, Redlands, California
LOUISA KOCH, Office of Education, National Oceanic and Atmospheric Association
TAMARA SHAPIRO LEDLEY, Center for Science Teaching and Learning, TERC, Cambridge, Massachusetts
MICHAEL TOWN, Redmond High School, Redmond, Oregon

MICHAEL A. FEDER, *Study Director* (until February 2011)
SHERRIE FORREST, *Study Director* (since March 2011)
MARTIN STORKSDIECK, *Director, Board on Science Education*
PAUL G. STERN, *Senior Scholar, Board on Environmental Change and Society*
REBECCA KRONE, *Program Associate*
ANTHONY BROWN, *Senior Program Assistant*

BOARD ON SCIENCE EDUCATION

Acknowledgments

This workshop summary is based on discussions at a workshop convened on August 31 and September 1, 2011, by the Board on Science Education (BOSE), in collaboration with the Board on Environmental Change and Society (BECS) and the Division on Earth and Life Studies (DELS), of the National Research Council (NRC). I would like to thank my colleagues who served on the steering committee, each of whom brought deep and varied expertise to the process of planning the workshop. The members of the steering committee developed the agenda, identified and selected presenters and paper authors, and facilitated discussion throughout the workshop. Although they did not participate in writing this report, this summary reflects the insightful comments of the steering committee members, presenters, and many of the workshop participants. The workshop was generously supported by the National Science Foundation as part of the NRC Roundtable on Climate Change Education.

This report has been prepared by the workshop rapporteur as a factual summary of what occurred at the workshop. The planning committee's role was limited to planning and convening the workshop. The views contained in the report are those of individual workshop participants and do not necessarily represent the views of all workshop participants, the planning committee, or the NRC.

This workshop summary has been reviewed in draft form by individuals chosen for their diverse perspectives and technical expertise, in accordance with procedures approved by the Report Review Committee of NRC. The purpose of this independent review is to provide candid and critical comments that will assist the institution in making its published report as sound as possible and to ensure that the report meets institutional standards for objectivity, evidence, and responsiveness to the charge. The review comments and draft manuscript remain confidential to protect the integrity of the process. On behalf of the NRC, I thank the other individuals who joined me in reviewing this report: David E. Blockstein, executive secretary, Council of Environmental Deans and Directors, senior scientist, National Council for Science and the Environment; Betty Carvellas, teacher leader, Teacher Advisory Council, Colchester, Vermont; Joe E. Heimlich, professor and specialist, Center of Science and Industry, and School of Environment and Natural Resources, and, Environmental Science Graduate Program, Ohio State University, and senior research associate, Institute for Learning Innovation; Karen Hollweg, principal investigator, A Framework for Assessing Environmental Literacy, North American Association for Environmental Education; Rajul Pandya, director, Spark, University Corporation for Atmospheric Research, Boulder, Colorado; and Bora Simmons, director, National Project for Excellence in Environmental Education, North American Association for Environmental Education.

Although the reviewers listed above provided many constructive comments and suggestions, they were not asked to endorse the content of the report nor did they see the final draft of the report before its release. The review of this report was overseen by Cary I. Sneider, Center for Science Education, Portland State University. Appointed by the NRC, he was responsible for making certain that an independent examination of this report was carried out in accordance with institutional procedures and that all review comments were carefully considered. Responsibility for the final content of this report rests entirely with the author and the institution.

I am grateful for the leadership and support of Martin Storksdieck, director of the Board on Science Education and the Roundtable on Climate Education, and Paul Stern, senior scholar, Board on Environmental Change and Society. I would like to thank Sherrie Forrest, workshop study director and associate program officer of BOSE and the Ocean Studies Board, for her support in planning and organizing the workshop. I would also give my sincerest thanks to Alexandra Beatty for serving as rapporteur. Anthony Brown deserves our thanks for his invaluable logistical support throughout the project and at the meeting, and Rebecca Krone for her support throughout the workshop planning and on the days of the workshop. Finally, I would like to acknowledge the support of the

Executive and the Reports Office of the NRC Division of Behavioral and Social Sciences and Education. Through their efforts we were able to bring together a diverse group of presenters and participants who shared their experiences and expertise at this workshop. The insights that we gained are summarized in this report.

Finally, I would like to extend a special thanks to the speakers and participants whose insightful contributions to the issue of climate change education have been recorded in this workshop summary.

Charles W. "Andy" Anderson, *Chair*
Steering Committee on Climate Change Education
in Formal Settings, K-14

Contents

1 Introduction 1

2 Student Understanding of Climate Change 11

3 Science Education Standards and Climate Change 23

4 Teacher Understanding and Preparation 37

5 Innovations at the High School and College Levels 49

6 Closing Discussion: Major Messages and Parting Thoughts 63

References 71

Appendixes

A Workshop Agenda and List of Participants 75
B Climate Change Education Roundtable 85
C Biographical Sketches of Presenters, Steering Committee Members, and Staff 87

1

Introduction

> Climate change is occurring, is very likely caused by human activities, and poses significant risks for a broad range of human and natural systems. Each additional ton of greenhouse gases emitted commits us to further change and greater risks. In the judgment of the Committee on America's Climate Choices, the environmental, economic, and humanitarian risks of climate change indicate a pressing need for substantial action to limit the magnitude of climate change and to prepare to adapt to its impacts. (National Research Council, 2011a, p. 1)

A principal message from the recent National Research Council report *America's Climate Choices*, this brief summary of how climate change will shape many aspects of life in the foreseeable future emphasizes the vital importance of preparation for these changes. The report points to the importance of formal and informal education in supporting the public's understanding of those challenges climate change will bring, and in preparing current and future generations to act to limit the magnitude of climate change and respond to those challenges. Recognizing both the urgency and the difficulty of climate change education, the National Research Council, with support from the National Science Foundation, formed the Climate Change Education Roundtable. The roundtable brings together federal agency representatives with diverse experts and practitioners in the physical and natural sciences, social sciences, learning sciences, environmental education, education policy, extension education and outreach, resource management, and public policy to engage in discussion and explore educational strategies for addressing climate change. Roundtable chair James Mahoney noted that the importance of the roundtable lies in its broad focus on climate change education, including formal education from kindergarten through college, public understanding, and the means to develop in decision makers, from local, state, and federal government officials to business owners, the capacity to address climate change issues.

The roundtable, which is overseen by the Board on Science Education, the Board on Environmental Change and Society, and the Division on Earth and Life Studies, was charged to hold two workshops to survey the landscape of climate change education. The first explored the goals for climate change education for various target audiences (National Research Council, 2011b). The second workshop, which is the focus of this summary, was held on August 31 and September 1, 2011, and focused on the teaching and learning of climate change and climate science in formal education settings, from kindergarten through the first two years of college (K-14). This workshop, based on an already articulated need to teach climate change education, provided a forum for discussion of the evidence from research and practice regarding:

- how climate change is currently taught in school;
- how best to teach climate change in K-14 settings;
- what factors impede the teaching of climate change in schools; and
- innovations in K-14 climate change education.

The goal of the workshop was to raise and explore complex questions around climate change education, and to address the current status of climate change education in grades K-14 of the formal education system by facilitating discussion between expert researchers and practitioners in complementary fields, such as education policy, teacher professional development, learning and cognitive science, K-12 and higher education administration, instructional design, curriculum development, and climate science. In an effort to provide a common frame for the workshop participants, the steering committee based the initial assumptions about climate change on the recent National Research Council (2010) report *Advancing the Science of Climate Change* that climate change is already occurring, is based largely on human activities, and is supported by multiple lines of scientific evidence. Beyond this initial assumption, the workshop did not discuss, nor intend to explore, the science of climate change or related climate issues, but rather confined the discussions to informing the issues around teaching climate change in formal school settings, K-14.

To explore these topics, the steering committee structured the workshop to provide ample opportunity for discussion among expert researchers and practitioners across the K-14 formal education system. This report

BOX 1-1
Workshop Statement of Task

An ad hoc committee will plan and conduct a public workshop that will address the current status of climate change education within grades K-14 of the formal education system. The workshop will feature invited presentations and discussion. It will provide an opportunity for discussion between expert researchers and practitioners in complementary fields, such as education policy, teacher professional development, learning and cognitive science, climate change, K-12 and higher education administration, instructional design, and curriculum development. Discussions at the workshop will focus on identifying how the issue of climate change is currently taught in school; what research indicates about how best to teach climate change in K-14 settings; what factors impede teaching climate change in schools; and how to best articulate the connection between climate change education in K-12 and higher education. The committee will develop the agenda topics, select and invite speakers and discussants, and moderate the discussion.

summarizes the workshop's presentations and discussions. Box 1-1 presents the workshop statement of task.

CONTEXT

America's Climate Choices (National Research Council, 2011a) describes key issues the nation faces in responding to climate change and developing strategies for mitigation and adaptation, noted Charles W. (Andy) Anderson (Michigan State University) in opening the workshop. The report articulates two challenges for the formal education system: to prepare scientists, leaders, and practitioners with the needed expertise to address climate change issues, and also to prepare *all* citizens to become informed decision makers. The report proposes that decisions about mitigation and adaptation be viewed in a framework of iterative risk management. That is, Anderson explained, the optimal response to climate change would be "an ongoing process of identifying risks and response options, advancing a portfolio of actions that emphasize risk reductions and are robust across a range of possible futures, and revising choices related to the climate over time to take advantage of new knowledge." The report does not call for a commitment to some particular course of action, Anderson noted. Instead, it asks for a commitment to understanding the implications of different courses of action and choosing in a deliberative way among them.

America's Climate Choices identifies key elements of an effective national response, Anderson explained, one of which is to develop institutions and processes that ensure that pertinent information is collected and that links scientific and technical analysis with public deliberation and decision making. Deliberation and decision making are critical to effective responses to climate change, Anderson emphasized, and thus it is essential to prepare all citizens to understand the risks of both action and inaction and to engage in effective deliberation about all available choices.

The vital importance of an informed citizenry is illustrated, Anderson noted, in data presented at the Roundtable on Climate Change Education's first workshop, on the diversity of beliefs people hold about climate change. In a series of studies that examined how the American public responds to climate change information, researchers categorized the public into "Six Americas"[1]: the six basic response categories are "dismissive," "doubtful," "disengaged," "cautious," "concerned," and "alarmed." "It's disturbing," Anderson observed, "that between 2008 and 2010 public opinion shifted away from concerned toward dismissive."

An even more important issue demonstrated in these studies, Anderson stressed is that public understanding of factual issues related to climate change is distinctly limited (National Research Council, 2011a). He stated that "as the evidence mounts, the controversy [about climate change] is inevitably going to die down" but noted that public deliberation becomes difficult or impossible when individuals choose their own facts, as well as their opinions and values, to interpret information. Anderson observed that in the future the controversy may shift from whether climate change is occurring to what actions need to be taken to address it and pointed to the need to prepare today's children for a future in which the basic facts of climate change are no longer controversial, and the consequences are real.

CLIMATE CHANGE EDUCATION FOR A CHANGING WORLD

Two assumptions underlie the way many people approach the topic of climate change education, noted Daniel Edelson (National Geographic Society) in the keynote presentation: one is that such education would begin with components of the science education curriculum, and the other is that much of it would concern climate science and the dynamics of climate change. Although he supports the idea that young people need to learn about climate processes and the ethics of anthropogenic climate change, he has a different view of the best way to conceive of climate change education.

To explain his perspective, he listed educational goals that are widely

[1]See http://environment.yale.edu/climate/ [January 2012].

shared in the climate change education community. Every graduate of the K-14 education system should understand

- Fundamental processes that influence climate, at scales ranging from local to regional to global.
- Natural variability and natural cycles in climate.
- Human impact on the climate—that is, how the growth in human population and technology has made it possible for human activity to change climate patterns at various scales.
- How changes in climate can and do influence physical systems, ecosystems, and society.
- Why the scientific community is now convinced that anthropogenic climate change is under way.
- What the range of effects of climate change is and how likely various different scenarios of climate change are under different conditions.

These ideas, he suggested, make up the common ground for educators who may have different perspectives on how to approach climate change education. People may draw different conclusions based on how this material should be covered in schools, but these basic ideas are the foundation for informed debate and decision making and thus provide a reasonable definition of climate literacy. He believes strongly that "the future of society and earth's ecosystems do hang in the balance, and . . . will depend, ultimately on our success in preparing the next generation to make good decisions [related to] climate change."

However, Edelson does not believe it is necessary, or even desirable, to create a major component of school curricula focused on climate science and climate change to achieve this goal. His primary reason is conceptual: he believes that most of what one needs to understand in order to be climate-literate has nothing to do with climate in particular, but rather is covered by the fundamentals of earth systems science. Edelson pointed to a report published by the National Aeronautics and Space Administration (NASA) in the 1980s that played a leading role in recasting the geosciences as the study of interconnected processes that cut across the traditional disciplines of geology, oceanography, climatology, hydrology, ecology, and others and also wove in the role of human systems (NASA Advisory Council, 1988). This approach was illustrated in a figure that came to be known as the Bretherton diagram (see Figure 1-1) after its author Francis Bretherton of the University of Wisconsin–Madison.

The Bretherton report was published at a time when climate change research was "gaining traction" and was very influential in the scientific community, Edelson noted. The report was intended as a guide to

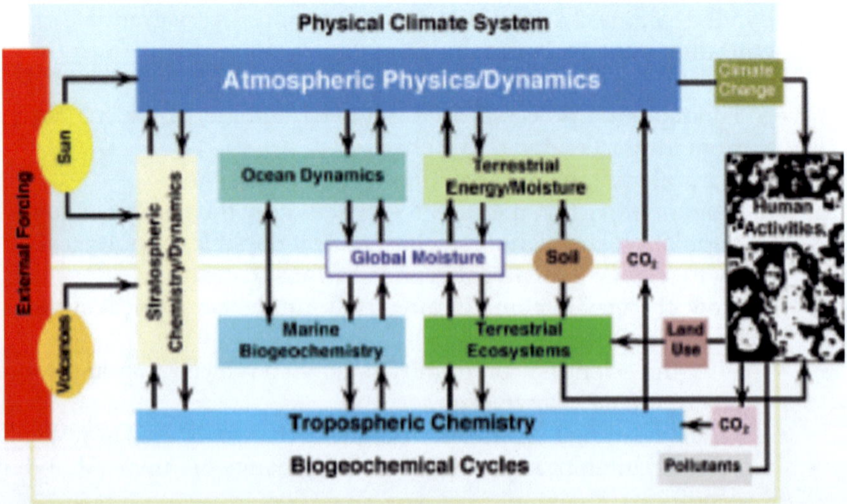

FIGURE 1-1 A diagram of the earth system proposed by the Earth System Sciences Committee, chaired by Francis Bretherton, which published the seminal report *Earth System Science: A Program for Global Change*.
SOURCE: Adapted from NASA Advisory Council (1988).

earth science research, and it put forth two critical principles: (1) that all earth processes and phenomena can be understood as dynamic systems that transform or transport matter and energy in accordance with the laws of chemistry and physics and (2) that all of these earth systems are interconnected, so that no system can be understood in isolation from any other. These were "paradigmatic breakthroughs," Edelson observed, which are now uncontroversial in the earth science community and which distinguished that field from other science disciplines that tend to study systems in isolation.

More recently this approach has been diluted, Edelson explained, by a tendency in the field of earth science education for people to advocate for particular spheres. For example, a framework for ocean literacy, developed collaboratively by the National Oceanic and Atmospheric Administration (NOAA) and others, described a sequential development of understanding of ocean science (National Oceanic and Atmospheric Administration, 2004). This framework was well constructed, in Edelson's view, but treats ocean literacy in isolation, "without regard to any other important literacies in the earth system sciences." He suggested that this framework follows the first principle of earth systems science—the dynamic systems principle—but sets aside the principle of interconnectedness. Some of those concerned with other areas of earth systems science perceived the framework as a challenge, Edelson explained, and responded with similar documents defining the knowledge sequence for these other areas.

The problem, for Edelson, is that "it's not possible to teach climate science on its own so that people can understand it." It only makes sense as part of an integrated, holistic education in earth systems science, although specific learning outcomes related to climate change would be part of that education. He commented that a shift in focus from climate change education to earth systems science education that encompasses key knowledge and understanding about climate change is necessary. Indeed, he suggested, "considering climate change education as a separate thing from overall earth system education is a waste of time."

Edelson thinks it important for concerned scientists and educators to "raise the banner of earth systems education and use climate change as one of the reasons that it is absolutely critical." To support this position, he cited a pragmatic example of why the climate-focused approach is not effective. In 2008, he pointed out, following a major effort by numerous groups of scientists and other stakeholders to educate the public about climate change, 72 percent of Americans believed that global warming was happening. By 2010, however, this figure had dropped to 50 percent, and the percentage of those who actively did not believe in global warming had doubled, from 17 to 36 percent (Borick, Lachapelle, and Rabe, 2011). The reason, Edelson proposed, is that the information the public had received was not connected to and embedded in larger knowledge structures. Thus, he explained, "as soon as somebody else comes along with a compelling argument that goes the other way, [people] cannot reconstruct the previous argument. They don't have the fundamental systemic understanding of what is going on."

Focusing on climate change in isolation places it in competition with other important education outcomes in a similar sphere, Edelson pointed out, such as study of the ocean, pollution, or even HIV/AIDS, poverty, or peace education. It is also important, in Edelson's view, to distinguish the study of climate change, which is an observable phenomenon, from the study of such fundamental, explanatory aspects of science as evolution, for example.

The exclusive focus on climate change also creates insider and outsider views, he added. Climate change is politically charged, and conflict over the new science standards and other education issues is inevitable. Such conflict would be much easier to overcome if the debate took place among people who already had achieved the basic literacy described above. The heat and energy of such conflicts can have long-lasting impacts, and Edelson expressed concern that conflict has the potential to derail the progress of the new science education standards that specifically include references to climate science and climate change education. It is important to remember, he noted, that "in politics you don't win because you are right, and you don't necessarily even win because you have a majority."

A better approach, for Edelson, is to pursue the goal of developing geo-literacy, or the capacity to make "big decisions," those that have big impacts (such as whether or not to drill for oil in a certain area) and that affect and are influenced by many processes and actors—such as formal democratic processes, advocacy, and public opinion. Edelson listed four components of geo-literacy:

1. reasoning about dynamic earth systems (physical, ecological, social, and technological);
2. reasoning about geographic relationships (relationships between places and systems and connections between places);
3. reasoning systematically about decisions; and
4. knowledge of specific systems and places at multiple scales.

This approach is intended to complement, not replace, other educational frameworks and standards, Edelson explained, but it provides a larger framework for integrating the key aspects of climate systems with those of human systems. It is also applicable not only to K-12 or undergraduate natural and physical science education, but also to the social and behavioral sciences curricula. It focuses on reasoning and decision making, with an emphasis on place and geographic decision making. "With the decline in geography education and social studies," he noted, "teaching of geographic reasoning and geospatial thinking has almost disappeared." Systematic approaches to decision making are not taught anywhere in the curriculum, Edelson noted, but people need "to be able to evaluate evidence, project consequences, weigh options and trade-offs, and use their values to make well reasoned decisions."

Adding climate change education on top of all that is already in the curriculum, in Edelson's view, is likely to yield a situation in which "some teachers do a great job with it, some teachers don't understand it or don't believe it and don't do it at all, and a lot of people will try to squeeze it in amongst a bunch of other competing priorities." Instead, he suggested, it would be possible to work backward from an understanding of the tasks that young people will face when they leave school and establish educational priorities that will truly prepare them. For example, Edelson emphasized, "students should be able to make well-reasoned decisions about purchases of cars and major appliances that take into account environmental impact, including climate change. They should be able to evaluate arguments for and against particular energy policies, and be able to communicate their own arguments to their elected representatives if they choose."

DISCUSSION

Participants had questions and comments about Edelson's proposed approach. Roberta Johnson (National Earth Science Teachers Association) asked whether Edelson was targeting a straw man, because, in practice, there are currently a very small number of courses being taught that focus on climate change. She commented that "if you are lucky enough to have an earth science class or an earth systems class, you might be doing a unit on climate change in the context of everything else you are doing." Edelson agreed, stressing that there should be both short and long-term strategies. In the short term, he said, it is important to work within the current system and integrate climate change into the classroom whenever possible. The long-term strategy he advocates is to address the fundamental problem with earth systems understanding and to treat climate change in the context of both earth and human systems.

Cathy Middlecamp (University of Wisconsin–Madison) questioned where biology, chemistry, and physics fit into Edelson's approach of fostering geo-literacy, and Ted Willard (American Association for the Advancement of Science) followed with a question about why focusing on geo-literacy is a better strategy than working across all science disciplines. Edelson explained that his approach is intended as a high-level organizational structure that should apply to all sciences in the curricula, pointing out that "we should be designing our education around what we want people to be able to do."

Nancy Songer (University of Michigan) asked whether it will be possible to take a top-down approach to developing the next generation of science education standards in a way that fuses together the content and practices for all of the sciences—natural, physical, and social—so that priorities can be set through a single concerted effort. Edelson reiterated that significantly altering the current standards effort "would be counterproductive" but stressed that the education community should already be focusing on the next generation of standards beyond the current efforts. He observed that a major breakthrough could come when coordinated standards for disciplines within social studies, which currently do not exist, are developed. Rather than eliminating boundaries between disciplines, Edelson suggested that connections across disciplines should be made well in advance (of articulating standards), so that the next generation of standards is better coordinated and integrated. "I am arguing for a better developed top-level structure, a way of connecting disciplines and making sure that when we set, say, physics priorities, they are connected to earth science priorities, math priorities, and social studies priorities, for that matter."

David Blockstein (National Council for Science and the Environment) pointed to the Bretherton diagram to ask how the definition of geo-literacy differed from that of ecology, noting that from his perspective it was a description of ecology. Edelson responded that the diagram is a marvel for just that reason—each community sees it as a model of their own discipline. Geo-literacy, he explained, goes beyond the traditional notion of earth systems to include the physical, natural, ecological, biological, social, and technological systems.

2

Student Understanding of Climate Change

Improving education about climate change begins with a clear picture of how students currently understand the issue, the quality of textbooks and curricular materials that are available for teaching about climate change, and how climate change is actually taught in classrooms and beyond. Eddie Boyes (University of Liverpool) described research on the mental models that students around the world have of global warming and climate change, Thomas Marcinkowski (Florida Institute of Technology) discussed climate literacy and pedagogy, and Frank Niepold (National Oceanic and Atmospheric Administration) discussed the nature and quality of available teaching materials.

MENTAL MODELS OF GLOBAL WARMING AND CLIMATE CHANGE

Climate change is a complex topic, noted Boyes: "Black body radiation, preferential absorption by the atmosphere at certain frequencies, the science of atmosphere and weather patterns, not to mention the stochastic nature of any predictions of climate fluctuations [based on evidence of] a warmer climate overall, and the risk assessments—these are very difficult issues." Good teachers are accustomed to making difficult ideas accessible to students by simplifying complex material and using models; and they also recognize the importance of understanding students' preexisting ideas, beliefs, and misconceptions.

A considerable amount of research has been conducted on people's ideas about science, Boyes added, including a new series of studies conducted at Yale University on public support for climate and energy change policies.[1] The research has shown that students have many misconceptions, Boyes observed (Shepardson et al., 2011). For example, many students confuse global warming with ozone depletion and believe that all pollutants contribute to all environmental problems. Many students also do not understand how global warming translates into climate fluctuations or why it does not simply make every place a bit warmer.

Boyes and his colleagues are currently conducting a study in 11 countries of people's beliefs and attitudes regarding climate change and possible actions that could be taken to mitigate it.[2] The researchers have sampled 12,627 students in grades 6 through 10 using a 32-item questionnaire that covers 16 issues, such as transport use, transport type, power generation, and selection of consumer durables. The questions focus on individuals' willingness to take particular actions and perceptions about how useful these actions would be.

Looking first at people's beliefs about the usefulness of various actions, Boyes described the overall approach and highlighted several questions as examples. Participants were asked questions, for example, such as "if people had smaller cars that used less gasoline, global warming would be reduced by ___" and given five options describing different levels of their perception of the effectiveness of the action: (1) by quite a lot, (2) by a fair amount, (3) by a small but useful amount, (4) by a very small amount (hardly noticeable), and (5) by nothing at all, really.

On many of the questions, responses varied across the countries. For example, responses to the above question on smaller cars resulted in the following: in South Korea, more than 70 percent of students chose the two most favorable answers: that smaller cars would help by "a fair amount" or "quite a lot"; in Brunei and the United Kingdom, approximately 50 percent of students chose these answers; in the United States, 64 percent of students chose these responses. Additional examples included questions that focused on such topics as how much people used their cars, alternative energy sources, and meat consumption (Boyes et al., 2011).

Questions assessing students' willingness to take these actions mirrored the questions about perceptions of effectiveness. For example, students were asked to choose a response to the question, "even if it was not as fast or luxurious, I would try to get a car that uses less gasoline." For this question, the five options were (1) I would definitely do it, (2) I would

[1]See http://environment.yale.edu/climate/ [January 2012], for a series of regularly updated studies.

[2]The countries are Australia, Brunei, Greece, India, Korea, Oman, Singapore, Spain, Turkey, the United Kingdom, and the United States.

almost certainly do it, (3) I would probably do it, (4) perhaps I would do it, and (5) I would probably not do it.

Here also the results were mixed, Boyes explained. For example, on the question of more efficient cars, students in India were most likely to say they would "definitely" or "almost certainly" try to get a car that uses less gasoline (nearly 70 percent), while 20 percent of students in the United Kingdom chose one of these two responses. In the United States, 38 percent of students said they would "definitely" or "almost certainly" get a car that uses less gasoline and 20 percent of them said they would use their cars less. Other questions focused on students' willingness to use public transportation, to pay for alternative energy sources, and to eat less meat (Boyes et al., 2011; Rodriguez et al., 2011).

The researchers also explored the connection between how potentially helpful respondents perceived an action to be and how willing they were to undertake that action (see Figures 2-1a and 2-1b). The five response items for the two question types were matched, Boyes noted, to make it easier to address the possible connections, and the researchers assigned codes to the five levels. Thus, the top responses ("the action would help quite a lot" and "I would definitely do it") were assigned a value of 1.00, the next responses ("help by a fair amount" and "I would almost certainly do it") were assigned 0.75, the middle options ("by a small but useful amount" and "I would probably do it") were coded as 0.50, the next ("by a very small amount" and "perhaps I would do it") as 0.25, and the lowest responses ("by nothing at all really" and "I would probably not do it") had a value of 0.00 (see Figure 2-1a). With these codes it was possible to create a scatterplot for the responses to both questions.

One might hope, Boyes observed, that there would be a linear relationship between perception of an action's efficacy and willingness to take that action, and the data did show a relationship between the two. Boyes also pointed out what he described as both "a natural willingness to act, up to a point," even for people who believe the action won't be helpful, and "a natural reluctance to act," even among people who believe the action might help a lot. These effects vary by question. For example, people are very willing to switch off unused appliances in their homes, but very unwilling to support nuclear power, even if they believe strongly in its usefulness to combat climate change. Workshop participants raised the question of what "willingness" really meant to respondents of the survey, noting that it could, at least in part, signify their recognition of what they ought to say or reflect the views of their parents or other influences. Boyes acknowledged the possibility that there would not be a perfect correlation between what students or others report being willing to do and what they later do, but pointed out that there are practical limits to the possibilities for assessing people's intentions.

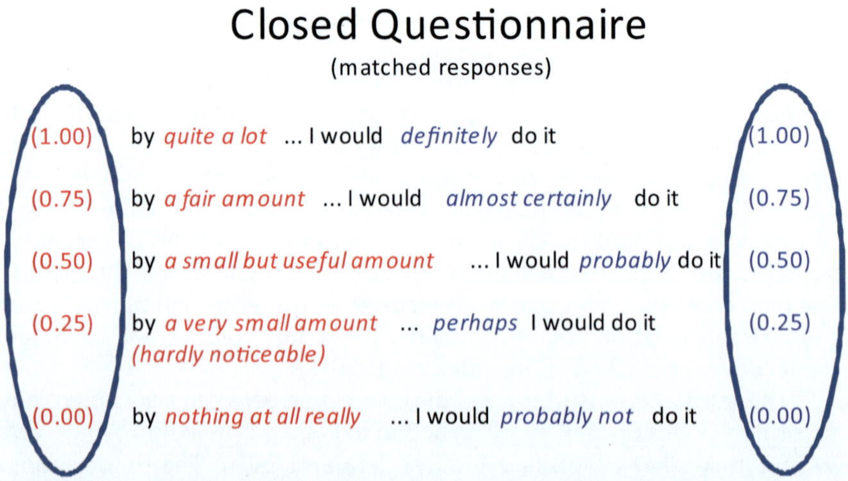

FIGURE 2-1a Codes of matched responses that connect the two question types: individual's willingness to act (responses on the right) and their belief in the usefulness of the action (responses on the left).
SOURCE: Boyes et al. (2011).

What the plot highlights (see Figure 2-1b), Boyes explained, is the zone in which education is likely to be most beneficial. The plot for each question can point to the areas between natural willingness (where education is not particularly needed) and natural reluctance (where education is unlikely to make much difference), where additional knowledge can increase people's sense of how effective an action might be and thus the likelihood of their taking that action. Such data, Boyes concluded, can be an important guide "as we try to educate rounded young people and make them critically thinking adults who can make decisions for themselves." He closed with a perspective on education articulated by Martin Luther King in 1948 (King, 1948):

> The function of education is to teach one to think intensively and to think critically. But education which stops with efficiency may prove the greatest menace to society. The most dangerous criminal may be the man gifted with reason, but with no morals.

CLIMATE LITERACY AND CLIMATE PEDAGOGY

Researchers have also explored the climate literacy of K-14 students, and Marcinkowski described the results of a secondary analysis of four national-level assessments of environmental literacy that have been conducted in Israel, Korea, Turkey, and the United States (Chu, Shin, and Lee, 2005; Shin et al., 2005; McBeth et al., 2008, 2010; Negev et al., 2008; Tal et

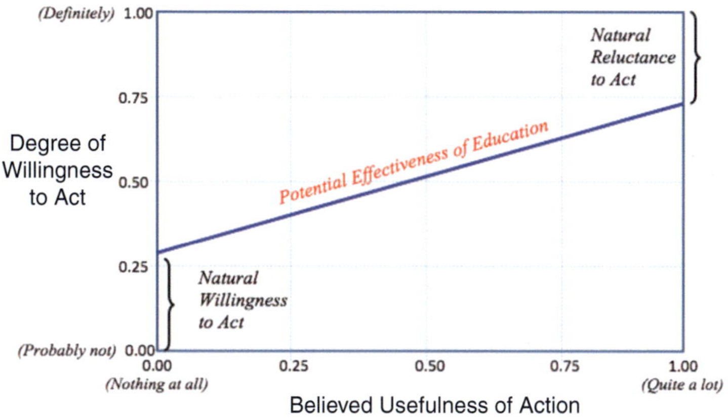

FIGURE 2-1b Graph demonstrating the relationship between individuals' willingness to act and their belief in the usefulness of the action. A willingness to act despite a person's belief that the action will not be useful to address a specific goal represents "a natural willingness." Conversely, there is a "natural reluctance" to act even when the action is perceived as highly useful.
SOURCE: Boyes et al. (2011).

al., 2008; Erdogan, 2009; Erdogan and Ok, 2011). Marcinkowski and his colleagues reviewed the results of these assessments to gain a sense of overall levels of climate literacy in each of the countries, how the countries compare in this respect, and how climate literacy varied across different ages. The analysis focused on data selected to represent climate literacy as defined by the U.S. Global Research Program/Climate Science Program, (2009, p. 3). According to that definition, a climate-literate person

- understands the essential principles of Earth's climate system;
- knows how to assess scientifically credible information about climate;
- communicates about climate and climate change in a meaningful way; and
- can make informed and responsible decisions with regard to actions that may affect climate change.

Climate literacy differs in some ways from more general scientific literacy, Marcinkowski noted, because behavior plays a more important part in it, in the sense that the actions and choices that result from climate literacy are "where the rubber hits the road." Table 2-1 shows the overlapping but not identical nature of scientific, environmental, and climate literacy.

TABLE 2-1 Domains of Scientific, Environmental, and Climate Literacy

Domains	Knowledge	Dispositions	Skills/ Competencies	Behavior
Scientific	X	X	X	
Environmental	X	X	X	X
Climate	X		X	X

SOURCE: Adapted from Marcinkowski (2011).

TABLE 2-2 Grade Levels and Content Covered in Four Assessments

	Israel	Korea	Turkey	United States
Grade levels	6, 12	3, 7, 11	5	6, 8
Knowledge				
Ecological knowledge	X	X	X	X
Environmental knowledge	X	X	X	
Affective dispositions				
Environmental sensitivity	X	X	X	X
Environmental feelings	X			X
Environmental attitudes	X	X	X	
Personal responsibility	X	X		
Locus of control/efficacy	X	X		
Verbal commitment/willingness	X	X	X	X
Cognitive skills/competencies		X	X	X
Behavior	X	X	X	X

SOURCE: Marcinkowski (2011).

The four assessments analyzed covered various grade levels and various components of the domain, as shown in Table 2-2.

To carry out the analysis, Marcinkowski and his colleagues had the three assessments that were conducted in other languages translated into English, and then they matched the items, along with items from the U.S. study, to key features of the climate literacy framework that is based on the American Association for the Advancement of Science's (AAAS') Project 2061 *Benchmarks for Science Literacy*.[3] The items that addressed aspects of climate literacy were a small subset (109 items in four assessments) of the battery of items included in the overall study, Marcinkowski noted, but the researchers charted the results for each of the components of the

[3]See http://www.project2061.org/publications/bsl/ [January 2012].

domain for which there were items by country and grade level. They reviewed these items to identify those that were the same or nearly the same across countries and grade levels, focusing on the results of those items. For example, they found that results of the Israeli, Korean, and U.S. assessments indicated that students' disposition scores were higher or more positive than their behavior scores, regardless of grade level. The researchers also found differences across both countries and grade levels, such as with the level of students' knowledge about different aspects of climate change, which suggest the need to take sociocultural and developmental factors into consideration when teaching about climate change (see Marcinkowski et al., 2011, for detailed analyses).

Marcinkowski and his colleagues also reviewed research in environmental education conducted between 1972 and 2005, on which he made a few observations (Marcinkowski, 2010). The researchers included both experimental work and other studies, some of which were not quantitative in nature, and searched for work focused on teaching and learning of natural history and ecology, environmental issue investigation, participation in service or action, and other approaches. While not exhaustive, this review of the research, in Marcinkowski's view, supports the proposition that different instructional approaches or programs tend to emphasize different outcomes. It is important to consider, for example, whether the desired outcome is increased conceptual knowledge or improved decision making when assessing the best approaches to climate education. In his view, what is most important is to remember that "what we are talking about is not playing on a team—we are creating a league." In other words, climate, ocean, environmental, and earth systems education can all play a valuable role, as can the social studies curricula. "There are a lot of players out there with strong communities and networks that have been and will continue to contribute to climate literacy. We don't need labels that are going to disenfranchise anybody—the challenge is to actively engage these constituencies."

TEACHING MATERIALS FOR CLIMATE CHANGE EDUCATION

A starting point for thinking about the materials used to teach about climate for Niepold is that "we don't have a century; we don't have half a century. We should have been moving on this three decades ago." Teaching about climate is happening today, he observed, and there are many materials in use, so it is essential to focus simultaneously on short-term goals as well as those that will take more time to reach, such as recasting the climate-related curriculum.

The climate literacy framework provides a sound basis, Niepold noted, because it incorporates the goal that students, learners, and citizens should be prepared to make informed and responsible decisions with regard to actions that affect climate. This goal poses a major challenge to the education system, he added, and will require a more comprehensive focus and integrative approach than exist in most of the current climate education resources, programs, and textbooks or curricula. At present, in Niepold's view, most resources are fragmented, and the topic of climate is not a priority in education systems. However, he stated that recent activities, such as the climate literacy framework, have begun to influence the development of resources, standards, and professional development materials and programs.

The climate literacy framework was the basis for a new means of evaluating the quality of climate education materials, he added. Funded by the National Science Foundation, the Climate Literacy and Energy Awareness Network (CLEAN)[4] is a collaboration among science and academic organizations[5] that was developed to identify and review online resources for teaching about climate science, climate change, and energy awareness, primarily those used in grades 6 through 16 and in informal science education. The resources are reviewed for scientific and pedagogical quality and are annotated by the reviewers. Resources are also aligned with the National Science Education Standards, the AAAS *Benchmarks for Science Literacy*, and the *Excellence in Environmental Education Guidelines for Learning* of the North American Association for Environmental Education.

CLEAN is designed to help educators locate excellent materials, although Niepold acknowledged, "excellence is a subjective thing." Nevertheless, CLEAN has helped to expand the framework within which climate must be understood, in part by including energy use in its purview, he explained. Thus far, the CLEAN researchers, in partnership with NOAA's Climate Program Office, have found what they classified as "excellent" resources that address over 15 percent of the learning goals presented in Project 2061's *Atlas for Science Literacy*[6]; an excellent resource or activity being defined as, above all, one that is scientifically accurate.[7] That is, the sources are accurate and trustworthy, the material is accurate and current, and there are proper citations and references. Excellent learning activities are also aligned with the identified climate and energy concepts, topics, and educational standards; they provide pedagogical scaffolding or teaching tips or are presented in such a way that educa-

[4]See http://cleanet.org/ [June 2012] for more information.

[5]TERC, Inc., the Science Education Resource Center (SERC) at Carleton College, the Cooperative Institute for Research in Environmental Science (CIRES) at the University of Colorado at Boulder, the National Oceanic and Atmospheric Administration (NOAA), the National Science Digital Library (NSDL), and the Colorado School of Mines.

[6]See http://www.project2061.org/publications/atlas/ [June 2012].

[7]See http://cleanet.org/clean/about/review.html [June 2012].

tors can develop their own instructional strategies; and they are easily accessible online and affordable. The CLEAN team has reviewed only materials that are available for free on the Internet, Niepold noted, adding that reviewing textbooks would be beyond the program's current capacity. One challenge, he cautioned, is that the science is moving so quickly that material can go out of date. However, Niepold stressed, the process benefits from the multiple layers of review and the materials are updated as new ones emerge.

The developers of the CLEAN database have also used the collected information to conduct some analyses. They found, for example, that approximately 16 percent of K-12 curricula, across all subjects, related in some way to climate literacy (including but not limited to climate change), which Niepold views as quite a large number (unpublished data, based on CLEAN and NOAA analysis using the NSDL Strand Map Server tool, the AAAS Benchmarks alignment to the climate literacy concepts). Among the 1,123 benchmarks for K-12, for example, they found 191 direct and related alignments to climate literacy concepts that spanned primary to high school grades across a wide range of curricular topics.

The team had several observations about the available resources, Niepold noted. In general, the resources they reviewed can be narrowly focused and their quality uneven. Some important areas, such as adaptation to climate change, are missing or thinly covered, and for others the resources lack interactive features likely to engage learners. Excellent resources can be difficult and time-consuming to locate on the Internet. He concluded that there is a need for better coordination so that effective practices can be more widely shared across disciplines, and that there is much more potential for integrated learning. Many of the resources are organized around a climate problem, he noted, and skillful balance will be needed "between gloom and doom on the one hand and inadequate strategies on the other."

Niepold emphasized the integrated, cross-disciplinary aspects of the study of climate and climate change. The ideas addressed in these curricula, such as consideration of the consequences of resource shortages, he noted, are complicated and represent rich topics that might be addressed in science, geography, or history class, for example. Such multilevel, multidisciplinary ideas, he added, are challenging to teach because teachers may address the subject matter in different ways that are not aligned with each other. He closed with a reminder that the guiding principle for informed climate decisions in the U.S. Global Change Research Program's climate literacy guide is that "humans can take action to reduce climate change and its impact" and that this principle sets the stage for the other principles, which are oriented toward "making informed and responsible decisions."

DISCUSSION

Participants provided a variety of comments and questions. Jay Labov (National Research Council) asked how children's beliefs about climate change develop and grow and how they are influenced by parents or other caregivers. He noted the polarization of opinion on climate change and observed that people's perspectives are often correlated with political affiliations. Boyes pointed to evidence that demonstrates clear relationships between sociocultural and political relationships and people's feelings about climate change in various countries, noting for example, that living in a democratic society seems to affect people's views. Niepold added that teachers and other educators do get challenged on this topic by a range of people, not just parents, and also noted that there is some evidence that young people find their information from sources other than their parents.

James Mahoney asked if there was evidence about whether students who form their own opinions apply them only to themselves or extend them to their communities—families, friends, other peer groups. He also wondered how effective education is at influencing the community. Marcinkowski responded that there are individual studies that suggest a positive influence from students "talking around the table" to their parents about what they learn in school. He also pointed to a study that followed students who were exposed to environmental issues in grade 8 and then provided no further intervening instruction. After two years without reinforcement, their level of skill and participation had decreased. On the positive side, students who wanted more experiences tended to find opportunities through organizations, such as scouts and other youth organizations that are not part of formal education.

Martin Storksdieck (National Research Council) asked Boyes if he has observed changes in students' understanding of climate change and global warming during his research over the past 20 years, wondering whether it has become more sophisticated, or whether people still struggle with the same issues. Boyes responded that students' understanding has matured since 2000. For example, there is less confusion about what the ozone hole is and how it relates to global warming. In addition, he noted there has been improvement in student learning over time from the lower to the higher grades, so that students do have a greater understanding of climate change in grade 12 than in grade 4.

Daniel Edelson asked whether there is a solid understanding in the social sciences of the progression from dispositions and beliefs to behavior, as students become adult decision makers. Both Marcinkowski and Boyes have found few studies that tackle those questions. They pointed to the scarcity of longitudinal studies and the difficulty of determining developmentally appropriate questions to explore the issue. Marcinkowski also noted that measuring affective dispositions, such as young people's attitudes, is challenging because attitudes and dispositions tend to be fluid and are likely to change from year to year as students develop and progress.

Andy Anderson closed by noting some conclusions and questions raised in the discussion. "Children live both in and outside of school" and are influenced by their families and their communities, as well as by their schools. That point leads to questions about the proper scope of the school curriculum and how to establish priorities among the knowledge, dispositions, skills, and behaviors that come under the heading of climate literacy. "What is the role that formal schooling can and should play in preparing students to be responsible citizens?" he asked. "What is reasonable to expect schools to accomplish? Where do schools have leverage, and where do they have permission from parents to go?"

3

Science Education Standards and Climate Change

Standards play an important role in determining what gets taught about climate, noted moderator James Geringer (Environmental Systems Research Institute). They provide the vision for teaching and learning and also define the categories that need to be covered and how they should be addressed through curriculum, instruction, and professional development. Brian Reiser (Northwestern University) and Stephen Pruitt (Achieve, Inc.) described the way climate change is addressed in the new *A Framework for K-12 Science Education* developed by the National Research Council (NRC) (National Research Council, 2011c) and the *Next Generation Science Standards*, which are currently being developed under the leadership of Achieve, Inc., and are based on the NRC framework. Gilda Wheeler (Office of the Superintendant of Public Instruction, State of Washington) provided a state perspective on standards, and Pruitt discussed some of the challenges that arise in addressing controversial science issues.

ADDRESSING CLIMATE CHANGE IN THE NRC FRAMEWORK AND NEXT GENERATION SCIENCE STANDARDS

The newly released *A Framework for K-12 Science Education*, Reiser explained, is not a set of science education standards but, as its title suggests, a framework to provide a vision and guide for the design of standards. The framework builds on previous documents, such as the College Board's Advanced Placement redesign, the American Association for the

Advancement of Science's (AAAS') *Benchmarks for Science Literacy*, and the *National Science Education Standards* (National Research Council, 1996), which have guided K-12 science education for 15 years. The first application of the new framework has been in the design of the *Next Generation Science Standards*, which are intended to replace current state standards over the next few years. The framework, however, could be used by any other entity that wished to develop science education standards.

The Framework

A Framework for K-12 Science Education[1] (National Research Council, 2011c) has three interacting dimensions: practices, crosscutting ideas, and core ideas, Reiser explained. Its structure is based on the idea that anything students learn in science is in some way a "melding of these three things." The framework is "a step forward from prior standards," Reiser added, because it reflects new understanding of how students learn and the challenges of using standards to guide instruction.

The framework suggests that new standards focus on "fewer, clearer, and higher" goals. New standards, Reiser noted, should be organized around "core ideas—not a million of them, not one per page, but a small number of core ideas per discipline." But, he added, this is "easy to say and hard to do." The committee that developed the new framework defined core ideas as those that

- have *disciplinary significance*, meaning that they are seen as key organizing concepts by scholars in the relevant fields;
- are *generative*, in the sense that they provide key tools for understanding or investigating more complex ideas and solving problems;
- are *relevant to people's lives*, in that they relate to the interests and life experiences of students and are connected to societal or personal concerns; and
- are *usable from kindergarten through grade 12*, that is teachable and learnable across the grades at increasing degrees of depth and sophistication.

For example, in the life sciences one of the core ideas is "from molecules to organisms: structures and processes"; in the physical sciences, a core idea is matter and its interactions.

[1]For a report brief of *A Framework for K-12 Science Education*, see http://www7.nationalacademies.org/bose/Frameworks_Report_Brief.pdf [June 2012].

The committee also identified seven crosscutting ideas, which are similar to those articulated in other standards:

- patterns;
- cause and effect: mechanism and explanation;
- scale, proportion, and quantity;
- systems and system models;
- energy and matter: flows, cycles, and conservation;
- structure and function; and
- stability and change.

Some of these ideas may go by different names in different contexts, Reiser stressed, but they apply across disciplines. For example, he explained, whether one is exploring phenomena in biology or earth and space sciences, a good strategy is to figure out where energy is going and how it is changing form: "Once you realize that you can't create or destroy any of it for free, that is, a really powerful heuristic that you can use across all different kinds of scientific problems." Several of these ideas are particularly valuable in explaining and reasoning about climate change, he added.

A second way the framework is important, Reiser explained, is in its commitment to the idea that learning develops over time. This is not a controversial idea, he noted, but "unfortunately this is not the way our science education system is broadly implemented." In his view, current approaches typically take necessary prerequisites into account but do not focus on carefully building understanding over time. For students to learn complex explanatory ideas they must revisit core ideas in new contexts that force them to extend their understanding, and engage in tasks that force them to synthesize and apply ideas. Reiser pointed out that there is a deliberate commitment in the new framework to the articulation of "a story line about how ideas develop over time." Figure 3-1 illustrates the progression of one of the core ideas, the structure of matter.

A third way the framework is important, in Reiser's view, is in its recognition that teaching content is necessary but not sufficient. Prior standards have also focused on inquiry, he noted, but the new framework better articulates what it means and how to teach it. "We can't teach scientific ideas without engaging students in practice that involves making sense of the ideas, applying them, extending them, explaining data—even using arguments from evidence to evaluate the consequences of different possible decisions," he explained. Thus, the framework calls on standards developers to create performance expectations that integrate all of these elements. Standards based on this approach would not yield "chapter one on the scientific method," he observed. "You can't teach the scientific

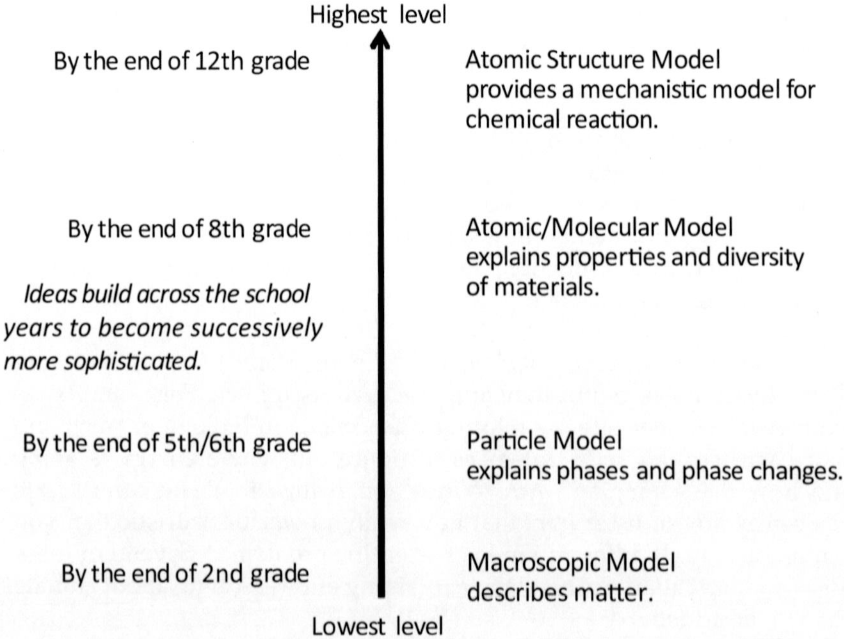

FIGURE 3-1 Progression for core idea: Structure of matter.
SOURCE: Reiser (2011).

method in the absence of reasoning about some scientific problem." Curricular materials developed in this way would not be purely expository narratives but would involve students in debating interpretations of data, making arguments to explain observed phenomena, and other scientific actions.

The scientific and engineering practices articulated in the framework are

- asking questions and defining problems;
- developing and using models;
- planning and carrying out investigations;
- analyzing and interpreting data;
- using mathematics and computational thinking;
- developing explanations and designing solutions;
- engaging in argument from evidence; and
- obtaining, evaluating, and communicating information.

Reiser emphasized that these are not separate chapters, but a vocabulary that should be in play constantly in the science classroom.

The framework provides guidance for how to develop standards, with an emphasis on performance expectations. For example, one core idea at grade 8 concerns conservation of energy and energy transfer: energy is transferred out of hotter regions or objects and into colder ones by the processes of conduction, convection, and radiation (National Research Council, 2011c). An associated practice would be developing and using models, and a suitable performance expectation for that grade level would be that "students create, defend, and communicate a model of the flow of energy and matter that explains how wind can occur" (see Figure 3-2) (National Research Council, 2011c). Figures 3-3a and 3-3b illustrate student models of understanding of energy conversion. Reiser stressed that it is less important that students memorize the correct terminology for labeling the diagram than that they understand the basic processes.

He closed with the observation that the approach to science education in general, and climate change in particular, that is articulated in the framework aligns well with the proposition put forward by Daniel Edelson's earlier presentation, that climate change should be treated as a topic within the larger context of earth systems science.

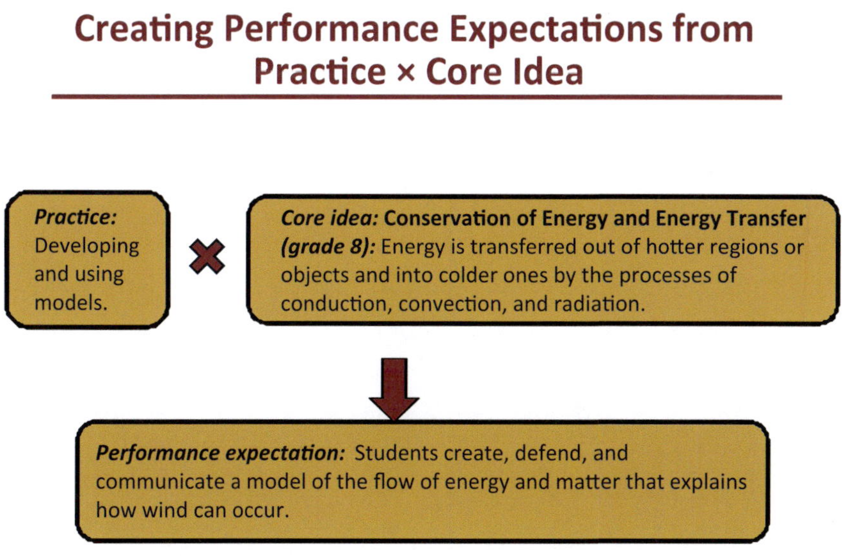

FIGURE 3-2 A performance expectation created from a core idea of *A Framework for K-12 Science Education* (PS3.B).
SOURCE: Reiser (2011).

FIGURE 3-3a Energy conversion and the practice of developing and using models. The student models show how energy is transferred from the sun to the surface, from the surface to the air by conduction (3-3a), and then how differences in temperature of the two regions can cause convection in the atmosphere, resulting in wind (3-3b).
SOURCE: Reiser (2011).

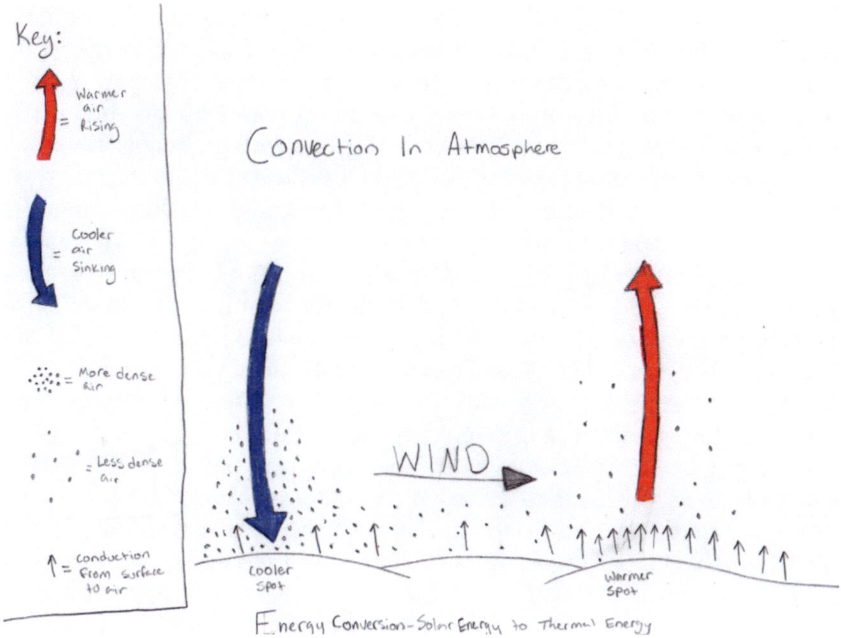

FIGURE 3-3b

The Standards

The *Next Generation Science Standards*[2] will faithfully reflect the framework's approach, Pruitt explained, and he provided an overview of the process to come. Although associated with the Common Core State Standards Initiative, the process of developing the *Next Generation Science Standards* differs in a few ways from the way standards were developed for reading and mathematics, he noted.[3] States were not asked to commit to adopting them before the standards were developed, as they were with the reading and mathematics standards. The developers wanted to "begin with the science," Pruitt explained, and collaborated with distinguished and internationally known scientists from relevant disciplines in defining the key knowledge, concepts, and skills in the NRC framework. This approach is expected to have the benefit of giving states that decide to adopt the standards a buffer if they face objections from some constituents over controversial issues.

The Möbius strip-like triangle that symbolizes the standards has three

[2]See http://www.nextgenscience.org.

[3]See http://www.corestandards.org/ [January 2012] for more information about the Common Core Standards Initiative.

sides to represent the core three primary elements of the framework that Reiser described: the core disciplinary areas, crosscutting concepts, and science and engineering practices. Three colors (orange, blue, and green) that represent these elements will be used throughout the document so that readers will be able to easily see how the elements are integrated. For Pruitt, the key improvement will be in the leanness of the document—it will not have long lists of facts and concepts for students to learn, memorize, and regurgitate when there is a test.

Pruitt acknowledged that a lot of work is needed to translate the framework into coherent standards that clearly identify student actions and that are easy for teachers and test developers to use. This phase of the process is being led by states, he explained, and several will be identified to lead the process and work intensively with the developers. The developers are a team of approximately 40, made up of K-12 educators, researchers, practicing scientists, and engineers; some members of the team have expertise in the education of students with disabilities and English language learners. There is also a much larger stakeholder committee (which includes approximately 700 individuals and organizations) that will provide feedback. The process will also entail two opportunities for public review as well as review by the science community. Members of the NRC framework committee will stay involved to help ensure that the standards are well aligned with the framework.

The team expects to have the standards ready by spring of 2013, and there is a plan to guide states in implementing the standards once they adopt them. A strategic advisory team will guide this process and assist states in engaging the business community and building coalitions in support of the standards.

A STATE PERSPECTIVE: WASHINGTON

"Climate change education isn't really about saving the planet, it's about saving humanity," Wheeler observed. She has found this approach to be very valuable because it engages people in a way that traditional environmental education sometimes does not, even in a state that has emphasized teaching and learning about the environment for a long time. Washington is unusual in having had a requirement since the early 1990s, she explained, that instruction about "conservation, natural resources, and the environment shall be provided at all grade levels in an interdisciplinary manner" (WAC 392-410-115, 2009). Recently, Washington has begun to structure that education around the theme of sustainability, and it has identified three primary benefits: a healthy environment, a vibrant economy, and an equitable society. This framing has made it easier to

BOX 3-1
Standards for Environmental and Sustainability Education, Washington State

Standard: Ecological, Social, and Environmental Systems
Students develop knowledge of the interconnections and interdependence of ecological, social, and economic systems. They demonstrate understanding of how the health of these systems determines the sustainability of natural and human communities at local, regional, national, and global levels.

Standard: The Natural and Built Environment
Students engage in inquiry and systems thinking and use information gained through learning experiences in, about, and for the environment to understand the structure, components, and processes of natural and human-built environments.

Standard: Sustainability and Civic Responsibility
Students develop and apply the knowledge, perspective, and vision, skills, and habits of mind necessary to make personal and collective decisions and take actions that promote sustainability.

SOURCE: Office of the Superintendent of Public Instruction for Washington State (2009).

involve a range of disciplines and content areas, more teachers, and more diverse stakeholders in the process.

This focus on sustainability is now reflected in the state's standards[4] for K-12 education (Office of the Superintendent of Public Instruction for Washington State, 2009a). Three standards are meant to be integrated into both the science and the social studies curricula, which address (1) ecological, social, and economic systems; (2) the natural and built environment; and (3) sustainability and civic responsibility[5] (see Box 3-1). Wheeler noted that there is actually only one place in the sustainability standards where climate change is explicitly mentioned: elaboration under Standard Two refers to "learning that is about the environment and environmental issues (e.g., loss of biodiversity, climate change, and water quality)" (Office of the Superintendent of Public Instruction for Washington State, 2009b). The state's science standards are similar to the plan for the *Next Generation Science Standards* in being structured around crosscutting concepts; in Washington's case, the concepts are systems, application,

[4]See http://www.k12.wa.us/Science/pubdocs/WAScienceStandards.PDF [January 2012].

[5]See http://www.k12.wa.us/EnvironmentSustainability/pubdocs/ESEStandards.PDF [January 2012].

and inquiry. Climate change is mentioned as an example in several places under systems and application, rather than treated as a unit.

The same is true for the social studies standards, in which climate change appears more frequently, Wheeler noted. For example, the standards for geography address "the United States' ability to meet the challenge of global climate change," the history standards include "ways to address global climate change that promote environmental sustainability and economic growth in the developing world," and a standard for skills suggests "small-group dialogue where each student presents two or more possible resolutions to the threat of climate change."

Washington has also defined a "specialty endorsement" for teachers who develop expertise in environmental and sustainability education. It covers core competencies in content knowledge, instructional methodology, and other professional competencies. Six teacher preparation programs in the state now allow teachers to earn this endorsement as part of a set of credentials. The state also offers a Career and Technical Education (CTE) course in sustainable design and technology and has developed an Environmental and Sustainability Literacy Plan,[6] in compliance with a provision of the federal No Child Left Behind Act.[7]

Wheeler closed with some data she collected in an informal survey of 94 teachers, of whom 90 percent were in Washington State. The survey focused on elementary, middle school, and high school teachers and targeted science teachers; 77 percent of teachers surveyed reported that they teach about climate change. The majority do so in the context of a science class, but a few mentioned doing so in a social studies, multi-subject, or mathematics class. For 60 percent, climate change is a subject that comes up occasionally, Wheeler reported. For almost a quarter of the teachers, the subject was treated as a unit within a course, and for a very few it was the focus of a full course. A quarter of the teachers reported spending 3 to 5 days on the subject in the course of a year, and 17 percent reported spending 2 to 3 weeks.

The teachers reported that, for resources and materials, they relied most heavily on videos, scientific articles, government websites, and curricula developed by private-sector organizations. Fewer rely heavily on textbooks or reports of the Intergovernmental Panel on Climate Change. Teachers cited as their greatest needs: instructional materials (78 percent), professional development (60 percent), links to content standards (56 percent), and support (political support, 41 percent; community/parent support, 40 percent; and administration support, 35 percent). The teach-

[6]See http://www.k12.wa.us/EnvironmentSustainability/pubdocs/WAESLPFinalJuly2011.pdf [January 2012].

[7]See http://www2.ed.gov/nclb/landing.jhtml [January 2012].

ers' comments reinforced their desire for greater support, Wheeler noted, and also highlighted ways that climate change can be addressed outside science and social studies classes. The comments also revealed that some of the teachers doubted the scientific consensus regarding the causes of global climate change.

In Wheeler's view, Washington's experience is an example of the importance of both national and state standards and policies. Although national policies provide both guidance and support, she concluded, states are critical for addressing what is needed in school districts and classrooms.

ADDRESSING CONTROVERSIAL SCIENCE ISSUES

"Whatever controversy occurs at the state level, most of the time it is actually an outcropping of what is happening in local communities," noted Pruitt. Moreover, the story is often a bit more complicated than what is reported in the media. For example, if a state school chief agrees to remove the word "evolution" from a set of standards, it may be the end result of a struggle to prevent the state legislature from outlawing the teaching of evolution outright or to allow alternative explanations to be taught. Such concerns generally come from parents and local communities. Pruitt pointed out that, in Georgia, for example, there was considerable debate when the state increased the rigor of its mathematics standards. As pass rates on state tests declined precipitously and students' grades fell, parents and others complained that the new standards were flawed.

Similar issues arose regarding the teaching of evolution in Georgia, he noted, but a strong constituency of community leaders, the business community, and the scientific community spoke out in support of the standards. The goal is for every science teacher in the state to be able to teach good science without fear of retribution, but he suggested, "that is not going to happen until we educate the full community." This is tricky, he added, because it is easy for policy makers and others to "come off sounding like zealots," and turn off the very people they are trying to reach.

The development process for the *Next Generation Science Standards* has been allotted 18 months, Pruitt added, because it is important to leave time to build understanding of what is in them and their importance. The key will be at the community level, he added, and he urged workshop participants to watch for opportunities to offer education at the local level.

DISCUSSION

Several aspects of standards and public education came up in discussion. Moderator Geringer stressed the importance of considering carefully the best ways to communicate to a broad audience. He cautioned against comparing climate change issues to other controversial science issues, noting for example that "if you mention that this is like evolution, immediately you have lost a lot of people." Terms make a difference too—"global warming" has already taken on a connotation for many people that limits its usefulness, and "anthropogenic," a word used in many reports about climate change "doesn't communicate in language people understand."

"You have to lay out a way forward," he added. If someone simply advocates that people stop using fossil fuels and ignore the potential impact on communities whose livelihood could be threatened, he explained, he or she "won't be communicating very well." Students and teachers need to be encouraged to think through possible solutions, he added.

A participant noted that it is common for people who are troubled by some aspects of a standards document to focus on the parts they are comfortable with and ignore others. In response, Pruitt suggested that the more such documents "tell a story" or offer an integrated narrative, the more difficult it will be to ignore parts of them. The states selected to lead the development of the *Next Generation Science Standards*, he added, have signed on to seriously consider adopting them, without the option of adopting only parts of them. Moreover, the standards are being designed to actually be adopted, rather than as a model for states to use in developing their own, as previous national standards were. "We are going to gift wrap a set of standards in a very rough economic time that states typically don't have a lot of money to do their own," he added.

BREAKOUT GROUP DISCUSSIONS

Participants were provided the opportunity to break into small groups to continue the discussion. Workshop participants had a choice of participating in breakout groups focused on one of two topics—the role of standards in climate change education or teacher preparation and understanding—based on their interest. Two groups of approximately 20-25 participants formed to discuss the topic of the role of standards. Each group was moderated by a steering committee member, and was also asked to identify a spokesperson to report back 1-3 main ideas during a plenary session at the end of the day. Four questions about the role of science education standards were presented as a starting point for the discussion:

1. What is the role of new science education standards and other frameworks (e.g., state environmental literacy plans or standards) in providing opportunities for or barriers to climate change education? How is the new framework similar to or different from current practices?
2. In addition to the areas identified in *A Framework for K-12 Science Education*, where should climate change education be covered in the curriculum?
3. In the translation of the framework to the standards, what are the opportunities to embed climate change literacy more broadly across disciplines?
4. What are the leverage points for incorporating climate change education into each level of education (elementary, middle, and secondary)?

The discussions were wide-ranging and provided opportunities for participants to exchange ideas and perspectives. A significant focus was the interdisciplinary nature of climate change. One representative from each group reported back, highlighting key points from their conversations, summarized below:

- There is a place for climate change education in most academic subjects. For example, mathematical modeling (mathematics); green technology (vocational education); communication and social discourse (reading, sciences, social studies); and visualization skills (arts) are all aspects of existing curricula that have a role in climate change education. There was some concern, however, that if treated only as a crosscutting concept and/or an important example of multiple scientific concepts, the topic of climate change itself could be lost. Making earth systems a core component of the curriculum, however, would result in the treatment of climate change as less voluntary. Some argued that if cross-disciplinary integration is to be truly meaningful, it will be necessary to reconsider the entire K-12 curriculum.
- Climate change instruction could be packaged as part of a curriculum on sustainability, which might both make it more personal for students and be easier to present in communities. It should be spiraled across the grades, with basics about data collection, graphing, and so forth starting in the early grades, while more complex social and political implications would be addressed at the higher grades.
- If climate change is to be taught effectively across disciplines, teachers in all subjects will need professional development to

boost their understanding and also to develop ways to integrate it into the curriculum. Standards and assessments will need to reflect this interdisciplinary goal, and new sorts of resources for teachers will be needed.

- It is not enough for a state to include climate change education in its academic goals. Not enough is known about how decisions are made at the district and classroom levels about what to include and how to present it.
- Standards provide a key support for teachers in dealing with controversial issues, such as climate change—for example, teachers can point to the standards when facing opposition by somebody who may not believe in climate change or may not agree with the subject matter that is being taught.

4

Teacher Understanding and Preparation

The science of climate change is very difficult to teach, observed moderator Tamara Ledley (TERC), not only because of political pressures but because the content itself is difficult to comprehend. Students have difficulty grasping the complex, interactive systems involved, and teachers sometimes struggle as well, she noted. Susan Buhr (University of Colorado at Boulder) and Roberta Johnson (National Earth Science Teachers Association) discussed aspects of teachers' knowledge, preparation, and practice.

TEACHERS' EXPERIENCES AND ATTITUDES

The controversy that has been a persistent factor in public opinion and policy related to climate change has also affected classroom teachers, observed Buhr. To illustrate her point, she drew on research on teacher practices and teacher learning: a descriptive study (Wise, 2010); a 2009 national-scale needs assessment (Lynds, 2009); an evaluation of a professional development workshop (Lynds, 2010); and survey results from a national study of teachers at the middle school through undergraduate levels (Hirabyashi, 2011) (see Figure 4-1).

The samples and methods used in these studies varied, but together they present a picture of teachers' experiences and approaches, Buhr explained. With regard to preparation, Lynds (2009) found that almost all the teachers in the national sample had engaged in self-directed learning of some sort to expand their understanding of climate change, using such

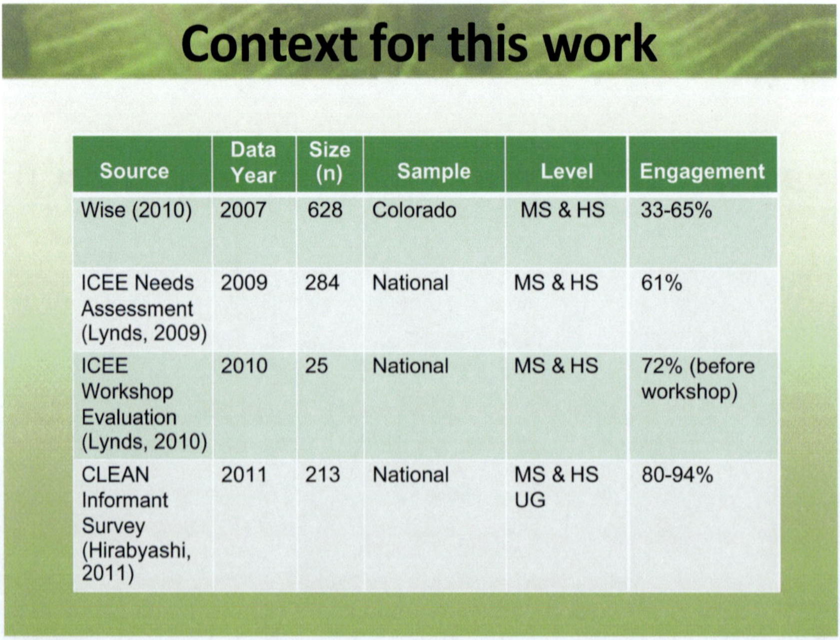

Context for this work

Source	Data Year	Size (n)	Sample	Level	Engagement
Wise (2010)	2007	628	Colorado	MS & HS	33-65%
ICEE Needs Assessment (Lynds, 2009)	2009	284	National	MS & HS	61%
ICEE Workshop Evaluation (Lynds, 2010)	2010	25	National	MS & HS	72% (before workshop)
CLEAN Informant Survey (Hirabyashi, 2011)	2011	213	National	MS & HS UG	80-94%

FIGURE 4-1 Four studies on teacher practices and teacher learning that were drawn on during the presentation. Samples vary by geographic scope, sample size, level of instruction, and degree of engagement in climate instruction, but themes and findings are consistent throughout.
SOURCE: Buhr (2011).

resources as websites, books, articles, television documentaries, and movies. Many also participated in short-duration learning experiences, such as workshops and conferences, but fewer have had sustained preparation in college or graduate school classes, and very few have had professional development in their own district focused on climate change.

The other three studies offered insights about teachers' content knowledge, suggesting that most feel comfortable teaching about earth systems but less so with climate topics (such as the greenhouse effect); emerging topics (such as considering the question of what will happen in a particular place as a result of climate change); and considering scientific evidence (how scientists know what they know). As a result, Buhr explained, many teachers are vulnerable to counterclaims from sources devoted to disproving that climate change is occurring or is caused by human activity, such as the documentary films, *The Great Global Warm-*

ing Swindle[1] or *Unstoppable Solar Cycles*.[2] Eighty-five percent of Colorado teachers in the study by Wise (2010), for example, reported that they support teaching both sides of the issue. Twenty-five percent of those who present both sides do so as an accommodation of students with different views, or as an opportunity to explore a controversy, but 50 percent left their reasoning unclear; 25 percent believe both sides are valid. Some of that 25 percent were confused or disturbed by "climate gate" (the public release of stolen e-mail correspondence among climate researchers that some viewed as casting doubt on certain data),[3] and others were actively committed to a "denialist" view, Buhr explained.

Teachers also reported on factors that obstruct their teaching about climate change, Buhr noted, the top three being (1) a real or perceived lack of alignment between climate change content and standards they are asked to follow, (2) their own lack of content knowledge, and (3) the beliefs of parents and students (Lynds, 2009; Wise, 2010; Hirabyashi, 2011). Teachers also noted interference by local school board members, as well as students and parents whose beliefs resulted in resistance to teaching about climate change or climate science. Teachers report using a range of strategies in response, including treating controversy as a teachable moment; working to integrate instruction about the climate throughout the curriculum; using inquiry-based pedagogy; inviting outside speakers, such as climate scientists, to expand the instruction; and integrating the search for solutions to specific climate problems into the curriculum. Buhr noted that the first four strategies are similar to those used by teachers who address controversies over evolution in the classroom.

Many teachers agree that climate change concepts should be taught not only in earth and environmental science classes but also in biology and social studies (Hirabyashi, 2011). However, in practice, teachers with biology degrees tend to state they are not well prepared to teach about the topic, Buhr noted. She pointed out that far more students take a biology class in high school than a geology or earth sciences class, stressing that biology classes are a key opportunity to reach more students. Teachers also devote relatively little time to climate change—in a survey of 213 educators from middle school through the undergraduate level, the majority of middle and high school teachers reported spending less

---

[1]Originally aired in the United Kingdom on Channel 4, March 8, 2007, the documentary film was directed by Martin Durkin.

[2]Further information available at http://heartland.org/policy-documents/unstoppable-solar-cycles-rethinking-global-warming [June 2012].

[3]Several investigations, including those conducted by the U.S. Department of Commerce's inspector general at the request of Senator James Inhofe, Pennsylvania State University, the InterAcademy Council, the National Research Council, and the British House of Commons, cleared the accused scientists of any wrongdoing.

than 25 percent of instructional time (as a formal topic) on climate topics (Hirabyashi, 2011). Many, however, do report integrating climate topics with other material; for example, 50 percent say they integrate climate concepts with discussion of societal implications or solutions, and 80 percent report that they have learned to incorporate discussion of solutions throughout their treatment of climate topics, so that students will not be overwhelmed by the disturbing nature of predicted consequences.

This research also demonstrates that professional development can affect teachers' views and that many report being surprised by the strength of the scientific evidence after a professional development experience, Buhr noted (Lynds, 2009). Teachers also are influenced by experiences in which administrators, colleagues, and others encourage them in their teaching about climate change and are less influenced by discouraging experiences (Wise, 2010). The most valuable professional development is that which is sustained; it is when experiences last for more than 80 hours, Buhr noted, that you start to see changes in practices. While teachers report having considerable opportunities for professional development about climate change, much of it is a week or less in duration (Hirabyashi, 2011).

Buhr concluded with the hope that new standards offer an opportunity for greater alignment and an increasing opportunity to build places for climate change instruction into the curricula in many subjects, as well as a chance to increase interest in and appreciation of climate change's importance to the public.

TEACHER PRACTICES AND CHALLENGES

Johnson drew on additional sources of data to expand the picture of current teacher practice and experience with respect to climate change. The National Earth Science Teachers Association (NESTA), she explained, has conducted several informal surveys of K-12 earth and space science educators and she focused on two of them. (Johnson, 2011). These surveys were not administered to a randomly selected sample, she cautioned, but were offered to interested teachers through dissemination of information about the survey. Sixty-two percent of the high school teachers who responded to the survey teach about climate change in their classes; 36 percent of middle school teachers, and approximately 15 percent of elementary teachers do so as well. At the high school level, the majority of teachers who cover climate change and who chose to respond to the survey were male, while the reverse is true at the elementary and middle school levels.

The survey data (based on a nonrandom sample of teachers) showed that climate change content is typically taught as a unit, in earth science and environmental science classes. The survey revealed that one big concern for earth and space science teachers is that these subjects are being deemphasized in many school districts. Seventeen percent reported that earth sciences were being dropped from their high school curriculum, 16 percent reported that it is being changed from a requirement to an elective, and 19 percent reported that it is being moved from high school to middle school. In many districts these courses are being disassembled, with portions of the content being integrated into other science courses. Johnson noted that budget cutbacks have affected many aspects of the curriculum but that the impact on earth and space science seems to be "disproportionate."

The surveys also explored the teachers' preparation in earth and space sciences and showed differences in the levels of preparation that female and male teachers have received. For example, male respondents were more likely to have taken college courses on the subject and female respondents were more likely to have received in-service professional development (see Figure 4-2). However, all respondents were more likely to have had in-service professional development on climate science and climate change than to have had college classes that covered the topic. Figure 4-3 shows what the respondents reported about the climate change topics they teach.

The NESTA surveys also asked questions about attitudes regarding climate change. In answer to an open-ended question about whether the responding teachers had encountered difficulty or pressure from students, parents, administrators, or community members about various topics, they were most likely to say that they had had such trouble with respect to their teaching about evolution (68 percent reported this). Forty-two percent reported experiencing pressure not to teach about climate change, although in terms of their own views, the vast majority of responding teachers reported that they believe global warming is happening.

Responding teachers were also asked to indicate whether they had experienced any of a long list of possible difficulties in teaching about climate change (examples include "climate change is too controversial to teach in my classroom"; "I don't know enough of the basic science behind climate change"; "I don't have access to a good textbook that covers climate change"; and "I don't have enough time to teach about climate change"). More than a third reported that they have been influenced to teach "both sides" of the issue. Of those, just 4 percent say they were required to do so and about 50 percent do so because they believe both perspectives are valid. Some of the many written comments from the responding teachers on this subject are shown in Box 4-1. Nearly half of the teachers reported some degree of increase in positive attitudes about the teaching of climate change in their own school; nearly one-third

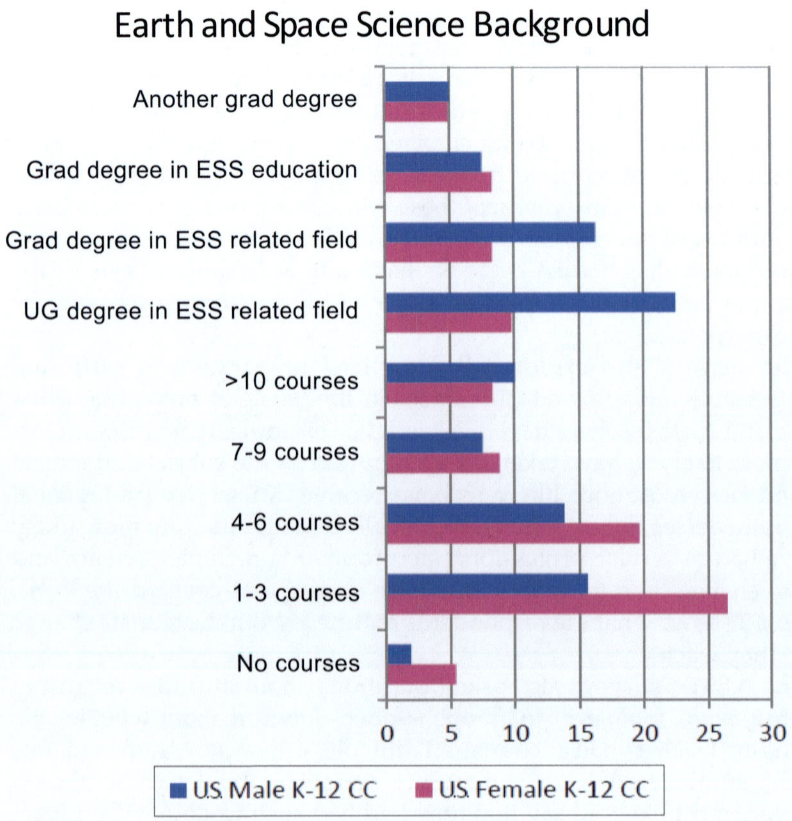

FIGURE 4-2 Percentage response by gender to the question, "Please indicate the amount of preparation you have had in earth and space science-related courses at the college/university level, including in-service professional development courses" (385 respondents).
SOURCE: Johnson (2011a).

reported no change in attitudes, and just under one-eighth reported some increase in negative attitudes.

Johnson had a few closing observations about the findings from these surveys, including a few points based on data she had not had time to present. First, in-service professional development appears to improve teachers' capacity to "represent what the science shows," she noted. Given evidence of differences in teachers' understanding by gender, region, and type and degree of preparation, as well as the finding that so many teachers believe they should teach "both sides," she stressed the potentially significant benefit of in-service professional development. Johnson

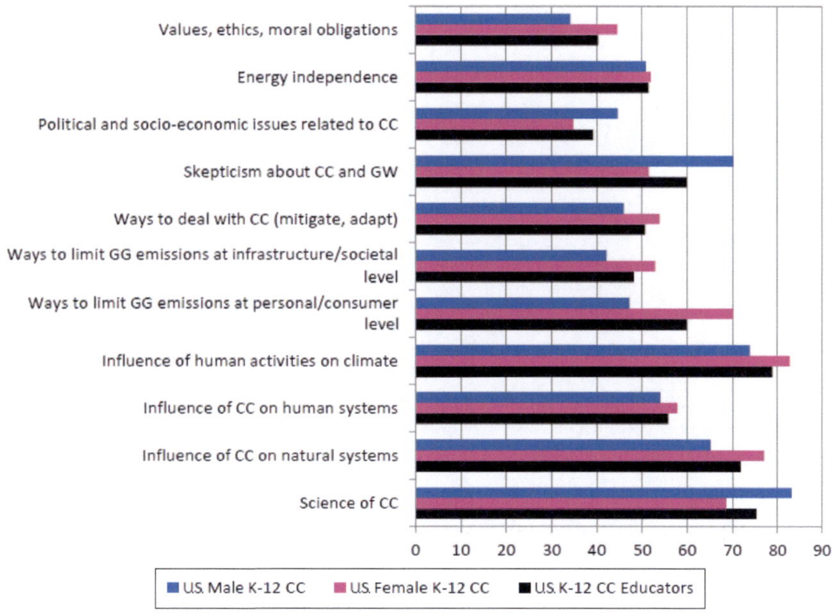

FIGURE 4-3 Percentage responses from U.S. K-12 climate change teachers to the question, "Which of the following major topics do you cover when teaching the subject? Mark all that apply." Topic order was randomized in the survey (372 respondents).
SOURCE: Johnson (2011a).

noted that "we need data-driven, inquiry- or discovery-based educational resources that are particularly well suited at the introductory level so that teachers that have a diversity of views in their classroom will have something they can do with the students, and can also dig in and discover it themselves."

Professional development needs to be intensive and long-term, she emphasized, but resources for teachers need to be flexible and realistic so that teachers can adapt them to the limited time they have for climate change instruction. "Students need to know," she commented, "that healthy skepticism and critical thinking are aspects of quality science, but that a refusal to accept overwhelming physical evidence is a sign of non-scientific thinking." Teachers need not be advocates, she concluded; "the evidence is overwhelming and the data can speak for itself."

REMARKS BY THE DISCUSSANT

There are about 1.6 million teachers of science in the United States, noted Francis Eberle (National Science Teachers Association) in offering

BOX 4-1
Sample Teachers' Comments About Challenges to Teaching Climate Change

"There was one parent who said he had a Ph.D. and said that he thinks global climate change is false. If I decided to teach it, he would come into the classroom and dispute it all in front of me. Instead of doing this, the school administrator just suggested that I not teach the information. I really found it to be a shame because it's something that is often found in the media that many students are interested in and want to learn more about. I don't want to tell them one way or another, but they need to know the facts from both sides without bias."

"I don't teach both sides. I only teach the science of it, not the politics. I talk about the politics, but there is no such thing as both sides unless discussing politics. Students then are given the opportunity to make choices on their own behalf after the science is explained. I don't think I have had any leave in disbelief in climate change, the science speaks for itself."

"I teach real science and bring in climate change whenever I want to show junk science or how a political agenda can slant people's honesty to produce false reports."

SOURCE: Johnson (2011a).

some context for the discussion, and they are a varied group. Some science teachers are exemplary and have outstanding skills and knowledge, others are not, and many are in between, so it is unwise to generalize about them. It is also important to remember, he added, how the science curriculum related to earth systems has changed over the years, having a variety of names (e.g., geology, earth and space sciences, earth systems science, environmental science) and a variety of subbranches (e.g., oceanography, meteorology, ocean and climate literacy). Thus, he suggested, students' experiences of the subject have varied far more over the past few decades than have their experiences with biology, chemistry, or physics.

Moreover, earth science courses are often electives, and their content may be determined by a teacher's knowledge and interests rather than a scope and sequence planned by the school or the district. The Advanced Placement exam drives the curriculum for many courses, he observed, but only a fraction of students take that course, and large numbers of students will only receive earth science education in a general science class. Right now, he observed, there are 20 states that do not require an earth science course for graduation.

Eberle highlighted some significant challenges that teachers face. In his view, science is not generally a high priority in states and districts. The requirements of the No Child Left Behind Act and other community priorities, he commented, have not put science first. Teacher preparation and professional development opportunities, he added, reflect this, which, in turn means that teachers may not have the opportunities to gain the knowledge of the science of climate change, and science more generally, that they need. "Systems are very complex. Understanding them, and the integration of systems—how many teachers have really learned about that?" he asked. If teachers, who are hired by communities, face opposition to the teaching of controversial material but have not had the preparation to clearly distinguish between scientific debates and political ones, he cautioned, "they are in a very awkward position."

The *Next Generation Science Standards* are thus "great news," in Eberle's view. Similarly, the current enthusiasm for science, technology, engineering, and mathematics (STEM) education should provide an opportunity to boost climate change education as well, he observed. At the same time, he added, the demographics of the teacher workforce are shifting, and as younger teachers come into the field, there is an opportunity to boost the general understanding of science and climate change issues. "Younger people are very optimistic," Eberle concluded. "They want to change the world. They want to do things that are exciting. They want to make a difference in their lives as they are growing up." This is the reason many go into science, and it may be a source of much-needed optimism about the challenge of climate change education.

DISCUSSION

"Which is the harder problem?" asked moderator Ledley, "helping teachers learn the concrete knowledge or dealing with their own skepticism about the topic?" One participant thought it possible to tackle both at the same time but noted the importance of avoiding the appearance of bias. This person added that "when I make a presentation on climate change and show a polar bear sitting on a dwindling iceberg, it probably looks biased to somebody who is outright against this. The data are overwhelming and the challenge is to make it accessible so that both teachers and students can effectively access it, probe it, analyze it for trends, and come to science-based conclusions."

Others thought that the most promising approach is to increase teachers' content knowledge and understanding of the nature of science. Several flagged the importance of observing the line between education and advocacy. "Teachers need to come down squarely on the side of the consensus view of climate science, but what is appropriate in the classroom is to equip students with the skills to make decisions about it, as opposed to telling them what their decision should be."

Right now, however, the significant fraction of teachers who are using climate change as an example of junk science, or teaching "both sides, have an amplifying factor equal to the number of students they are reaching," another noted. "We can't give up on them," this participant added, "and then there is that vast range of teachers who really don't know and are doing the best they can. They don't really have a good resource base. We can reach them with strong, data-driven resources that are nowhere near close to advocacy."

BREAKOUT GROUP DISCUSSIONS

Participants were provided the opportunity to break into small groups to continue the discussion. Workshop participants had a choice of participating in break out groups focused on one of two topics—the role of standards in climate change education, or teacher preparation and understanding—based on their interest. Two groups of approximately 20-25 workshop participants formed to discuss the topic of teacher preparation and understanding. Each group was moderated by a steering committee member, and was also asked to identify a spokesperson to report back 1-3 main ideas during a plenary session at the end of the day. Four questions focusing on teacher understanding and preparation were presented as a starting point for the discussion:

1. What types of pedagogical knowledge are needed to teach about climate change or climate science? How can one help teachers to obtain the knowledge they need to teach climate change comprehensively?
2. How can teachers and principals overcome skepticism about climate change and climate change education from, for example, parents or administrators?
3. What are strategies for finding appropriate curricular materials?
4. How can schools and districts organize themselves so that teachers are motivated to teach climate change?

One representative from each of the breakout groups reported back, highlighting key points from their conversations, which are summarized below.

- It is important to connect local changes to changes in the larger global system—to help students understand climate change.
- Teachers need to model the practices of science—to get students

out taking measurements and making observations and connecting what they are doing with what other researchers have been doing.

- It is useful to use misconceptions as a starting point for lessons. It is especially useful to focus on misconceptions that are not particularly controversial, as a prelude to conversation about more difficult ones.
- Teachers could benefit from opportunities to conduct research with practicing scientists so they can be directly exposed to scientific methods and data collection. They also need explicit instruction in how to interpret data. Having a single reliable source for professional development programs and materials related to climate and energy, such as the Climate Literacy and Energy Awareness Network (CLEAN) resources, would be very useful.
- Local professional development that explicitly links the K-14 curricula and establishes partnerships with scientists and researchers will help teachers become teacher-practitioners, rather than just teachers of science.
- The most useful professional development is long term, as opposed to "quick hits," and helps teachers develop better communication strategies.
- Stronger incentives would encourage teachers to pursue professional development related to climate change, such as explicit state endorsements for earth science and climate literacy, professional development credits, time off, and stipends for teachers who pursue continuing education.

5

Innovations at the High School and College Levels

Elective high school courses and postsecondary courses and programs differ from the required K-12 curriculum, noted Andy Anderson. The audience for upper level, elective courses is more limited, a situation that allows more freedom to innovate. Innovations developed in these contexts can also generate ideas for use in general K-12 classes. Karen Lionberger (College Board, Advanced Placement [AP] Program) described the AP Program in environmental science; LuAnne Thompson (University of Washington) described a dual-credit course for high school students offered by the University of Washington; Nicky Phear (University of Montana) described an interdisciplinary minor program on climate change offered at the University of Montana; and Matt Lappe (Alliance for Climate Education) described an award-winning climate change education program offered by a nonschool institution.

ADVANCED PLACEMENT ENVIRONMENTAL SCIENCE

AP environmental science is a relatively young program, compared with the other AP science programs, noted Lionberger. Started in 1998, the course is currently being redesigned to incorporate more student-centered, inquiry-based experimentation and instruction and to focus more on developing deep understanding, rather than covering a broad range of material.

Climate issues are explicitly addressed in three of the components of this one-semester course, she explained, each of which accounts for 10 to 15 percent of the total course content:

- earth systems and resources—the atmosphere (composition, structure, weather and climate, atmospheric circulation and the Coriolis effect, atmosphere-ocean interactions, El Niño/Southern Oscillation);
- the living world—natural ecosystem change (climate shifts, species movement, ecological succession); and
- global change—global warming (greenhouse gases and the greenhouse effect, impacts and consequences of global warming, reducing climate change, relevant laws and treaties).

However, she added, climate is a theme that runs through most of the course; it arises naturally in the context of many of the topics covered, such as energy and the formation of fossil fuels. The course also addresses the human impacts of global warming, such as the spread of diseases and increases in mosquito populations and ranges based on temperature changes. Students are asked to go beyond the environmental impacts and address such issues as the potential effects of environmental changes on society and economic conditions.

AP environmental science is one of the fastest-growing AP courses, Lionberger noted, averaging annual growth rates of 17 to 20 percent per year. However, just over 100,000 students took the course in 2010 (out of over 3 million students who took all AP courses in 2010). Students are generally excited about this course, she added, and the course design makes it easy for teachers to engage students through fieldwork, helping them see the material's relevance to their lives. One challenge for AP environmental science teachers is that few students come to the course having previously taken an earth systems or earth sciences course, whereas students in other AP science classes have often already taken a year's worth of coursework in the subject. Thus, AP environmental science introduces students to a wide range of material. Lionberger observed that, "it's an introduction to probably ten different majors that you could spend three years of intense coursework on. It really is a challenge to try to balance all of these topics and give students a broad but deep understanding."

This challenge is one of the focuses of the redesign of the course, Lionberger noted. "They are moving away from what, as a teacher, you'd call the march of topics into depth of understanding. Instead of spitting out information, students will be required to justify, argue, look at data rationally, and make an argument for their decisions," she explained. The program is also working on improving resources for teachers—not only for the AP teachers, but also for teachers at the elementary and middle grades, so that they can better prepare students for the AP course. The AP program is also focusing on incorporating 21st century skills into the coursework and has joined the Partnership for 21st Century Skills in this effort.[1]

Above all, she said, "we want the students to become interested in science. We want them to fuel the STEM field and to be excited about doing that coursework." The links with colleges and universities through which they validate the course content are invaluable in that regard, she added.

A DUAL-CREDIT COURSE ON CLIMATE

It is important for scientists to communicate with broader audiences about their research, but they face challenges in doing so, observed Thompson, a professor of oceanography and an adjunct professor of physics and atmospheric science at the University of Washington. Many are "kind of at a loss as to how to do it," she noted, and often fall back on "one-off" presentations. That problem is one of the reasons why the University of Washington applied for grant funding to develop a college-level course in climate science that could be offered in high schools. The primary goal was to connect research and education in climate science, she explained, and specifically to increase students' sense of the relevance of science, to create a sustainable means of outreach for University of Washington science faculty, and to bring the depth and interdisciplinary nature of climate science to high schools.

There are programs throughout the country, Thompson noted, that offer college coursework in high schools. Typically, high school teachers are trained by faculty at partner postsecondary institutions to teach university-level courses, and the university oversees both the content and the assessments students take. Students earn both high school and college credit. Such dual-credit courses allow them to experience college-level rigor in a familiar setting and also foster collaborative relationships between high schools and local colleges, Thompson explained.

In this case, the University of Washington establishes the curriculum activities, tests, and grading scales and selects the texts. The university offers such courses in English, foreign languages, calculus, geology, and other subjects. These courses provide an alternative to AP classes, Thompson commented, but also complement the AP program. Students do not earn credit through a single high-stakes test, but instead are evaluated as they would be in a college class and receive a grade that can go on their college transcript (AP exams yield credits that colleges may accept,

[1] See http://www.p21.org/events-aamp-news/press-releases/1001-new-members-for-p21 [December 2011] for more information.

but not college grades). The current grant (from the National Aeronautics and Space Administration) allows students to take the course without paying tuition, but ultimately, to sustain the course, students will have to pay tuition, she noted. Ten high school teachers have been trained to date, and the current grant should cover the training of an additional 10 by next year.

Climate and Climate Change is a five-credit course that is currently taught twice a year at the University of Washington as a lecture class. It is an introductory course for students not majoring in science and is required for those who wish to minor in climate studies. The university also offers a separate course on global warming that focuses on politics and sustainability, but the faculty chose to focus the option for high school students on climate science, Thompson noted. The course content overlaps significantly with the AP environmental science curriculum, covering:

- climate of the present—the global energy balance, atmospheric circulation, the role of oceans and ice in climate, the carbon cycle, and atmospheric composition;
- climate of the past, on time scales ranging from thousands to billions of years; and
- climate of the future—is the earth getting warmer? Why? How will the climate change over the next 100 years? Should people be concerned?

Building a teaching and learning community has been a primary goal for this project, Thompson emphasized, noting that many groups have gotten engaged and have benefited. Faculty and researchers in the departments of Atmospheric Sciences and Earth and Ocean Sciences are participating, which has helped improve their outreach to the community and ability to reach more students. Graduate students have been able to develop modules for the course and earn certification in the teaching of climate science. High school teachers have gained valuable experience through associated professional development opportunities. Undergraduate students have benefited from the recent establishment of a minor in climate science. High school students earn credit and a college grade, and gain college experience. In turn, the university hopes to encourage these students to matriculate and possibly boost enrollment in small departments, which helps the departments sustain themselves and grow (there are currently approximately 300 majors in the three sponsoring departments combined, compared with 1,300 in biology).

This degree of engagement is an important success of the program, Thompson observed. Highly qualified and engaged high school teachers have been recruited, and there is support at the principal, district, and state levels. The program has been established to be sustainable, but it still faces challenges. Although the program in some ways complements the AP, there are a limited number of students who opt for higher level courses, so some competition is inevitable. Staff at the University of Washington are working with AP staff to coordinate the two programs, Thompson added. There are differences in the styles of campus lecture-based classes and the more hands-on high school ones, and Thompson noted that she hopes the hands-on approach will influence the college classes.

Another challenge has been the recruitment of teachers from beyond the Seattle area to offer these classes. Identifying qualified teachers and arranging for them to travel to Seattle to receive the training has been difficult, she noted. Earth science does not have high status among high school faculty at present, and relatively few teachers have taken atmospheric science.

AN INTERDISCIPLINARY CLIMATE CHANGE MINOR

There are currently very few opportunities for focused interdisciplinary study of climate change at the undergraduate level, noted Phear. The University of Montana introduced such a program in 2009, inspired by the work of Nobel laureate and faculty member Steven Running, an Intergovernmental Panel on Climate Change author, who urged the university to recognize that global warming is likely to be the defining challenge for future generations and that students need to understand it and begin to respond. The university provost charged the faculty to design an undergraduate curriculum for the study of climate change that would be interdisciplinary and innovative and would emphasize problem solving and solutions.

The program was developed by a faculty task force representing many disciplines: geosciences, chemistry, geography, forestry and conservation, environmental studies, ethics/philosophy, communications, economics, political science, sociology, business technology, energy technology, and journalism. Also involved were representatives from the university's Wilderness Institute, through which students study wilderness and its stewardship through education, research, and service, and the Mansfield Center, which sponsors programs on Asian affairs, ethics, and public affairs.

This group's collaboration produced a plan for a minor in climate change studies that would be available to students from all majors. It requires them to take 21 credits, including a 3-credit interdisciplinary introductory course and 6 credits each in three areas: climate change science, climate change and society, and climate change solutions.

The science curriculum, Phear explained, introduces students to the basic processes by which the biosphere, atmosphere, hydrosphere, lithosphere, and cryosphere interact to produce and respond to climatic changes. The curriculum related to society provides students with the opportunity to evaluate the social, political, economic, and ethical dimensions of climate change at the local, national, and international levels. The solutions portion of the curriculum creates opportunities for students to study, develop, and implement solutions to climate change through internships and other applied coursework. All three are addressed in the introductory course (see Box 5-1) and in the minor courses from which the students select (see a sample in Box 5-2).

BOX 5-1
Introductory Course Topics, University of Montana Climate Change Studies Minor

Science

- Introduction and Principles of the Global Climate
- Global Climate Change Models
- Paleoclimatology
- Current Climate Change Trends
- Oceans
- The Cryosphere
- Climate Change and Forest Dynamics

Society

- Ethics and Climate Change Policy
- U.S. Climate Change Policy
- The Economics of Climate Change
- Adaptation and Geoengineering
- The Rise and Fall of Large Prehistoric Villages
- China and Climate Change
- Europe and Climate Change
- Psychology of Environmental Problems
- Communications and Climate Change

Solutions

- Climate Stabilization and the Wedge Solution
- Raising Energy Efficiency
- Energy Sources: Turning to Renewables
- Carbon Offsets
- Business Solutions to Climate Change
- Climate Action and Adaptation Plans
- The Power of Example: Individual and Collective Action

BOX 5-2
Climate Change Studies Minor Course Sample Offerings

Science
- Climate Change—Past and Future
- Weather and Climate
- Snow, Ice, and Climate
- Global Change
- Biogeochemical Cycles

Society
- Climate Change Ethics and Policy
- Sustainable Climate Policies: China and the USA
- International Environmental Economics and Climate Change
- Communication, Consumption, and Climate
- Environmental Sociology
- Psychology of Environmental Problems

Solutions
- Sustainable Business Practice
- Environmental Citizenship
- Introduction to Renewable Energy Systems
- Climate Change Internship
- Climate Change Practicum
- Climate Change Field Studies

Many of the courses offered are field-based, Phear emphasized. Students may research impacts of climate change in various locations. For example, they have the opportunity to do activities in Glacier National Park, including a survey of mountain goats and interviews with ranchers and people who live in forested areas to learn about the impacts of changes and adaptations. They also meet with doubters and decision makers. A study abroad course takes students to Vietnam to explore adaptation to rising sea level in the Mekong Delta. Every student in the minor program is required to complete an internship or take a course that includes an applied project involving campus initiatives, local businesses, government agencies, or nonprofit organizations. Recent examples have included work on sustainability programs for the university, work on biomass utilization and carbon accounting with the U.S. Forest Service, and work with the Pew Environment Group on development of a national climate policy campaign.

Phear highlighted two themes that are interwoven throughout the climate change studies minor. The program's developers defined a climate-literate person as one who can communicate effectively about climate and climate change; for that reason communication is the topic of specific courses, but also permeates all of the courses and experiences. Students are encouraged to have conversations and use such tools as blogs, symposiums, and wiki pages to learn about and consider perspectives different from their own and to engage in and deliberate about the issues raised by climate change.

The faculty members also stress the importance of fostering and engaging networks both in the university (across disciplines) and in the community. Phear observed that she views education as "an iterative, adaptive process in which students learn from faculty, carry those conversations across disciplines, and apply them on the ground." The 54 students currently signed up to complete the minor requirements, Phear noted, represent 23 different majors, which she identified as a key asset that will help engage students, even those who may not opt to complete the minor but will take the introductory and other courses. The students are very active on campus, she noted, and have also taken their enthusiasm to Washington, DC, and elsewhere. For her the question is not whether or not there should be climate change education, but what it should look like.

BRINGING CLIMATE CHANGE TO HIGH SCHOOL

The Alliance for Climate Education (ACE) is a nonprofit organization dedicated to educating high school students about the science behind climate change and inspiring them to do something about it, explained Lappe. Headquartered in Oakland, California, the group has educators in 10 cities and hopes to continue to grow. The group's primary approach is to present a multimedia based, energetic, and interactive school assembly that explains greenhouse gases and their sources in language and symbolism geared toward the culture of high school students and to follow up with support for students who are motivated to take action. In the 2 years since the group was founded, Lappe noted, assemblies have been presented to approximately 900,000 students in their schools. Approximately 180,000 of those students have signed up to stay connected through ACE's virtual network, and about 31,000 have become active members of environmental clubs associated with ACE. Another 1,100 students have participated in ACE training to support them in taking a leadership role in their communities. A small number of youth are recruited to take part in the presentations; they are referred to as Youth Representatives (Figure 5-1).

ACE was developed in response to the recognition that more traditional ways of trying to reach students have not yielded adequate results.

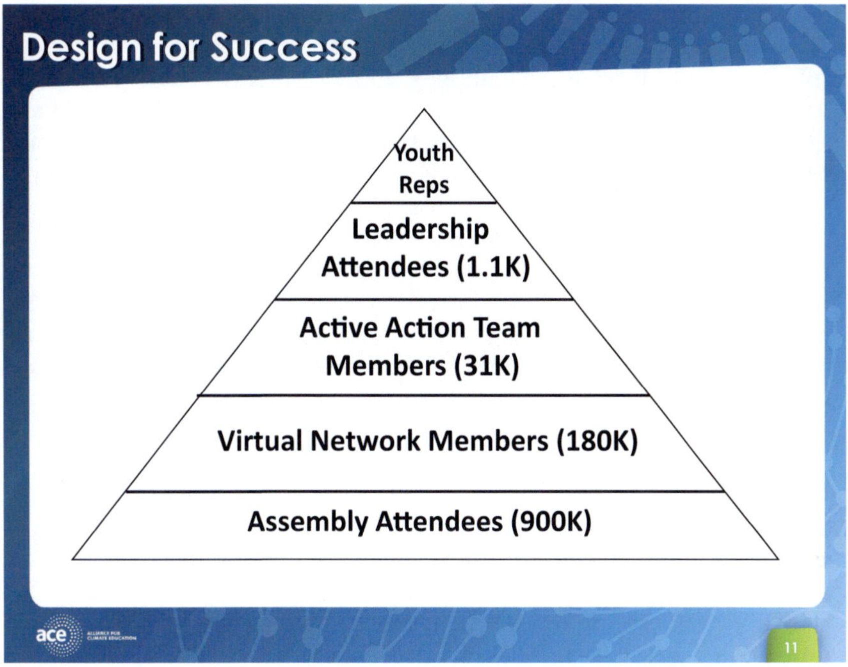

FIGURE 5-1 The levels of engagement of students targeted by Alliance for Climate Education assemblies.
SOURCE: Lappe (2011).

"It seems like climate scientists are always looking for that one magic piece of data that [will make people] fall to their knees, start crying, and realize that climate change is a serious issue," Lappe noted. "But, sadly, many people who have not had a background in the sciences are not receptive to that." The ACE presentation is based on peer-reviewed research and is overseen by a science advisory board composed of practicing scientists, Lappe explained. It also reflects sophisticated strategies for engaging students who may not be interested in the subject.

The various levels of engagement offered by the ACE program allow students to involve themselves to the degree that represents their interest and abilities. For example, the training might equip students with the tools to initiate an energy audit in their schools and sponsor a "biggest loser" energy challenge. ACE also has begun to focus on supporting teachers, assisting them in finding curriculum resources, and professional development opportunities and to find and develop professional networks.

ACE has also begun to measure the outcomes of its efforts, Lappe noted. For example, staff conducted a survey in spring 2011 of the knowledge, attitudes, and behaviors of students before and after an assembly presentation, an action team site visit, and leadership training. The survey collected data on about 300 students from 7 schools and 13 classrooms; ACE hopes to expand the survey to a total of 2,500 students. Preliminary results show an increase in the percentage of students who agree that "most scientists think global warming is happening" from 48 percent before to 59 percent after an assembly presentation, as well as an increase from 54 to 74 percent who agree that "the amount of carbon dioxide in the atmosphere today is higher than it has been over the past 800,000 years." After seeing an assembly presentation, students also reported feeling more confident in their ability to help start a project to reduce their own school's carbon footprint and in their ability to explain global warming to others.

REMARKS BY THE DISCUSSANTS

There are many programs, contests, and academic opportunities for students around the country, noted discussant Michael Town, an environmental science teacher who served as an Einstein fellow at the National Science Board. It is important, in his view, that climate change education proceeds on two tracks. There is a baseline, he explained: "We want all kids to know a certain amount about climate change," and standards are a key to meeting that goal. At the same time, "we have kids who are very passionate about this issue and want to get advanced training." There are 1.6 million seniors in high school and only about 100,000 of them take AP environmental science, he added, so all the programs out there have a vital role to play.

At his home school, Redmond High School in Washington state, the administration and faculty wanted "to be really ahead on environmental literacy," he explained. Almost half of the graduating students each year have taken AP environmental science, and 87 percent, on average, have passed the exam. The school also offers an independent science research course and the Cool School Challenge, an energy-auditing program initiated in 2007. The school has reduced its carbon footprint by 250,000 pounds and is saving the district $40,000 annually as a result. The school has also developed a vocational education class that covers green business and technology issues, such as the process of obtaining Leadership in Energy and Environmental Design (LEED) certification; green building techniques; and installation of solar, geothermal, and other alternative energy heating and cooling systems. Sixty-eight percent of the students are involved in the environmental club.

One of the challenges to maintaining these types of programs and courses, Town noted, is that there can be significant attrition when an individual who has taken the lead in developing programs leaves the school. For example, when he left for his one-year Einstein fellowship, enrollment in the AP classes declined, and the "Design and Sustainability" course was no longer offered. One issue, he added, is a shortage of teachers with the necessary experience and credentials. "When I got my degree," he explained, "I could not teach in a public school because environmental science is not an endorsed field like biology, chemistry, and physics." As a result, people with environmental credentials tend to pursue opportunities in informal education. This is a "really, really critical problem," he added: "That's why the need for professional development to help existing teachers retool is really critical." In his view, the standards are there to support the baseline education for all students. What is more challenging is a way to bring the opportunities for advanced study to all the students who are interested.

DISCUSSION

Participants had comments and questions. Moderator Louisa Koch asked about how learning about climate science and climate change might be related to the development of stewardship behavior. Lionberger noted the importance of passionate teachers, emphasizing that "kids care when their teachers care." In her view the learning is what sparks students to tackle these problems. Many of the teachers involved in environmental education sponsor clubs and other activities outside the classroom, Town added, and such activities can influence the culture of the school and help engage students who are not otherwise involved. Lappe emphasized the importance of making climate change something students want to get excited about by going beyond the walls of a "geeky" environmental club. Phear cautioned that activism is very inspiring for students but that some students who get involved may not actually understand the science of climate change and the policy issues. "They can't be very discerning about the effectiveness of their actions or why they might be turning people off," agreed a participant.

Bill Easterling (Pennsylvania State University) brought up the challenge of graduate education and the view that students who specialize in environmental or climate issues are still at a competitive disadvantage if they don't receive a degree in physics, chemistry, or geology. Pennsylvania State University offers dual-title degrees that are based in a discipline but emphasize climate change science. These degrees might lead to government jobs that offer the opportunity to shape policy. Thompson agreed and presented another model: at the University of Washington, students receive degrees in a discipline, such as atmospheric science, but also can get a certificate in specific area, such as communicating about climate science.

Sophia Gershman, a high school teacher from New Jersey with a doctorate in plasma physics, asked about teacher training, pointing out that most teachers were not exposed to "real science" as part of their undergraduate education. Lionberger agreed that it is essential to work with teachers to provide them with the skills to teach science with hands-on, inquiry-based practices, adding that even teachers with science degrees often do not teach this way in their classrooms. Thompson commented that in her experiences with professional development of high school teachers, there is a lot of potential to provide the needed tools and knowledge to help teachers with the science, but she emphasized that this takes a lot of time.

Christopher Crowson (National Environmental Education Foundation) asked whether the climate minor at the University of Montana has engaged with humanities and history departments. He was particularly interested in how history has shaped society's world view and values and how those connect with the current climate dilemma. Phear pointed out that the introductory course has a strong ethical component that focuses on the origins of social values and international comparisons. She also noted that science students who start out in this class tend to be uncomfortable when addressing these topics, whereas students who are more engaged with these topics tend to be uncomfortable with the science. It is important to address all of this material in one course, she noted, and that is a requirement for every student. Thompson added that scientists are also often uncomfortable engaging in conversations about ethics and the anthropological and historical underpinnings of people's values and often prefer to stick to scientific facts. In addition, faculty and administration in some disciplines may feel threatened by interdisciplinary programs that tend to attract students from a variety of disciplinary backgrounds.

Roundtable chair James Mahoney cautioned the education community about their use of the phrase "global warming." This phrase does not capture many important implications of climate change for the broader public, such as the potential for more intense hurricanes and erratic winter weather or ocean acidification, he observed. Lappe agreed, pointing to research regarding communication about climate science. Studies show that Americans relate the phrase "global warming" to polar bears and melting glaciers and do not see the connection with their lives. Lappe has found that it is important to link educational materials to local issues and students' real lives.

Carol Brewer (University of Montana) emphasized the transdisciplinary nature of the study of climate change and asked where biology fits into the conversation; there are a lot of consequences of global climate change that will take place in the biological realm, she noted. Thompson responded that there seems to be a strong focus in biology in the medical fields as students come out of high school. Lionberger added that although many high school biology teachers teach environmental science as part of their course, often they lack the necessary training in earth sciences. For her, this relates to the question of what can be done to prepare these teachers.

6

Closing Discussion: Major Messages and Parting Thoughts

Andy Anderson began the closing discussion with a reminder of the elements of an effective national response to climate change that were articulated in *America's Climate Choices*: enacting policies and programs that reduce risk by limiting the causes of climate change and reducing vulnerability to its impacts. In his view, what this means, is that, as a nation, the United States will need to consider when and how to forgo current consumption in the interest of future well-being.[1] *America's Climate Choices*, he noted, "is saying that we're in this for the long haul, and we need to think about how we are going to create and sustain the cultural changes that we will need in order to respond appropriately to climate change." He identified four primary challenges in providing the nation's youth with education to sustain this appropriate response, drawing on the presentations and discussion from throughout the workshop.

FOUR CHALLENGES

Challenge 1: Preparing for sustained efforts in the nation and in formal schooling. Brian Reiser discussed the importance of learning progressions as a basis for the framework for the new national standards, Anderson noted. This is key because "the things that we would like kids to learn they are not going to learn in a day or a week or a year. We have to think

[1]The conversation about the role of advocacy vs. education recurred during the conference and is addressed explicitly in Chapter 4.

about how we are going to have a sustained response in our schools to the learning issues that kids face." Yet earth science, the location in the framework for the new national standards on climate change, "is increasingly being driven out of high schools," Anderson noted. This raises the question of how the current structure of schools and curricula—and a teaching force that has not, in general, had the education necessary to teach effectively about climate change—will support the kind of sustained response that is needed.

Challenge 2: Finding the proper role of formal schooling in the national response. The workshop provided a variety of messages about the role of formal schooling, Anderson noted, which can have quite different implications. First, Daniel Edelson proposed a definition of geo-literacy that encompasses elements that fit within the traditional science curriculum (although he placed greater emphasis on human systems reasoning than the traditional science curriculum has), but also includes decision making, which has not had a place in the science curriculum. Thomas Marcinkowski offered another conception of what might go into the school curriculum, incorporating both traditional aspects (knowledge, cognitive skills, and competencies) and something new, in this case dispositions and behavior.

There was a lot of discussion, Anderson noted, about the degree to which these ideas present a significant challenge to science education as it is now configured. Core ideas are the guiding structures of current frameworks, but they are generally taught as a list rather than as an integrated set of ideas, he noted. Currently missing, in his view, are the crosscutting concepts and the related practices. Eddie Boyes, in turn, identified the "zone between the things nobody will do and the things everybody will do as the natural place where education can be effective—suggesting that that's what schools should focus on." While these ideas may converge, they do not at present suggest a complete consensus about the conception of or priorities for climate change education, Anderson remarked.

Anderson's own research has looked at how young people decide what the truth is about a situation they are considering, and he has found that they usually make use of personal and family knowledge, as well as ideas from media and popular culture. "They often make judgments about bias and self-interest in people and in organizations making the claims—they are often pretty perceptive about why you would not trust a particular person or group," he added. They rarely make use of knowledge they learned in school, he noted, or make explicit judgments about the scientific quality of evidence or arguments. Anderson finds this very troubling—and a real challenge to schools—noting that dialogue can take place only if people "understand when the people that they don't like and don't trust are still making a good argument." This point relates to Challenge 3.

Challenge 3: Thinking about the role of values in an issue about which passions run high. There are several reasons why conflicting values seem to be fueling passions in the discussion of climate change, in Anderson's view. First, there is a marked culture gap between scientists and the public. Climate scientists have been polled, he pointed out, about whether they agree that climate change is happening, and consistently more than 95 percent of them agree that it is. Yet only 13 percent of the public believes that more than 80 percent of scientists believe that global climate change is happening, Anderson observed. That 13 percent includes people from across the spectrum, he added: those who are alarmed or concerned about climate change as well as those who are dismissive.

"There's just this huge difference between what scientists believe and what the public believes that scientists believe," he commented. At the same time, many science teachers believe they should teach both sides of the climate change issue, although "in the opinion of scientists there are not two sides—there is a set of established scientific findings," he added.

This gap probably reflects differences in the ways scientists and non-scientists think about uncertainty, he suggested. Scientists have developed particular ways of dealing with uncertainty, he observed. They recognize that it is never entirely absent, and that they can never know that they have found absolute truth, but there are methods they use to reduce the uncertainty about the claims they make. For scientists, authority does not rest with individual people but stems from arguments based on evidence. "You don't trust somebody because he or she is smart or well positioned," he explained, "you say, 'what's the evidence?'" Scientists rely on rigor and research methods and on collective validation, peer review, and other ways of achieving consensus in the scientific community.

"These are values," noted Anderson. Scientists believe in and live by them, and face severe sanctions if they fail to do so. "That's why scientists trust reports like *America's Climate Choices* and others," he added. "They can't imagine the scientists who contribute to those reports violating those values in a systematic way."

These scientific values need to be taken into account, he added, in discussions of interdisciplinary climate change education. Many at the workshop advocated interdisciplinary approaches, but, he suggested, the disciplines are where those values reside. Scientists have developed their understanding of what rigor, evidence, and collective validation mean in the context of their fields of study. "We need to break down barriers," Anderson observed, "but if we abandon the standards and values that make science important, have we given up the baby with the bathwater?"

It is an important function of education, he added, to teach students to understand those values and recognize that scientists are passionate about them. "These are things that people argue about and get angry about and stake their careers on," he observed. The barriers between, for example, scientists and engineers remain, he added. The interdisciplinary work is not necessarily happening at the faculty level. "When and how do the engineers and the scientists and the people in the humanities come to talk about policies and strategies?" Anderson asked.

Challenge 4: Using what we learn from research and innovation. The programs described at the workshop do demonstrate many ways of engaging students across disciplines, Anderson noted. Curricula and programs were examples of ways to connect social, political, and economic issues to science. Others demonstrated ways of engaging many different people. The Alliance for Climate Education, for example, engages high school students who may be more interested in Lady Gaga than in climate science, and Redmond High School engages students through vocational courses focused on green technology and the building trades.

CLOSING DISCUSSION

During the final session of the workshop, presenters, panelists, steering committee members, and attendees discussed themes and issues that emerged. Workshop participants provided comments and engaged in discussion, followed by closing remarks by James Mahoney, Climate Change Education Roundtable chair, and Martin Storksdieck, director of the Board on Science Education and of the Climate Change Education Roundtable. This section is organized around the major themes that emerged during this discussion.

The Context of Climate Change Education

Several participants favored the idea that the study of climate change should be encompassed in a much broader earth systems curriculum and indeed should be a presence across the curriculum. James Geringer returned to the challenge Edelson had raised at the beginning of the workshop—should there be climate change education at all? From Geringer's perspective, it is not possible to teach climate change in isolation, because if it is isolated from an understanding of the bigger picture, people do not recognize how climate change can affect their lives. He emphasized that "if you understand the fundamental principles, such as natural variability, natural cycles, your understanding of climate change will come as a result." Geringer also highlighted the importance of teaching kids about scientific uncertainty and risk management, adding that "in many of our educational processes the students want to know is this yes or no.... It's only later in life they discover there are many answers." He added that students need to be taught to reason and ask questions in a way that helps them better understand the world.

Participants considered how climate change education could be positioned in K-14, wondering whether it is a great example of how to teach other disciplines in an integrated cross-disciplinary way, or a core science issue that should be taught on its own.

National and State Standards

Louisa Koch explained that she is very supportive of the new science framework, because it is very important to take a national approach to climate change education. The new framework and standards, one participant pointed out, build students' awareness and sophistication level gradually, "to the point where reasoning can come to bear and questioning can be pertinent." Michael Town stressed the role of states in promoting and sustaining environmental and climate change education. In Washington, he noted, they have implemented state standards and programs that support sustainability education—creating a position in the office of the superintendent, an endorsement for teachers who specialize in environmental sustainability, an environmental literacy plan, and classes that prepare noncollege-bound students with job skills for the green economy. He added that it will be very important to identify replicable and scalable programs that have successful track records and export them to other places around the country.

Communication

Participants also focused on communication issues. "We're overlooking a lot of people who are just uneasy with being told 'this is happening, so do something about it,'" noted Geringer. He suggested that one reason why so many people are confused about climate change is that they have not been taught how to reason and to ask questions. One participant noted that although there are not really two sides to the issue of climate change, scientists do have differences. For example, some focus on data from the past 30 years, whereas others look back 50 years, and these frames of reference may lead to moderately different assessments. It is important to teach students that there are different ways to assess and evaluate information that are equally consistent with the scientific method, this person added. Doing so, several others observed, will also be a way of "humanizing" scientists, helping people understand what it is they do and how they reach their conclusions.

Several participants emphasized the importance of developing trust between the scientific community and the public, noting the need for scientists to engage in more communication. Anderson elaborated on this point, noting that scientists are often seen only as doing individual work and have not been very successful in communicating that science is most often done as a community. Other participants emphasized the importance of telling stories that build a narrative so that students can connect with the issues. This can be accomplished through developing collaborations with such disciplines as art and history and approaching climate change through such issues as energy and health. As one participant suggested, "give kids a sense of where they're coming from, where they are, and where they can go and the real possibilities that apply out there."

Tamara Ledley emphasized the importance of creating bridges across different levels of learning and highlighted the importance of teaching students to communicate what they learn. She cited as an example a program at Dartmouth College that integrates learning at the high school and undergraduate levels and teaches students to bring what they learned to the broader community. She added that by reaching out to a wider audience, the program had the added benefit of making the information relevant to students at a personal level.

Interdisciplinary Nature

Lynn Elfner (Ohio Academy of Science) observed that an important goal is to prepare people to understand climate change issues so that they will take action, and that is why an interdisciplinary and multidisciplinary approach is important. Although it is essential to master a single discipline, he added, people need to understand how to use science for solving problems and making good decisions about real-world issues. Carol Brewer emphasized the need to "create partnerships to blaze the transdisciplinary trail and also to broaden one's own knowledge to be conversant beyond your individual expertise." She stressed that a starting point could be that "in our own classrooms we have to be brave enough, regardless of how our universities or schools are organized, to find a colleague to teach with." In searching for those partners, she emphasized, teachers can look to fill gaps in their own knowledge and improve their teaching.

Jill Karsten (National Science Foundation) pointed out that the current Climate Change Education Partnership Program at the National Science Foundation is designed to foster the development of partnerships among climate scientists, learning researchers, and education practitioners. The projects funded through this program also reach out to stakeholders in the communities in an effort to align the needs and efforts of the broader community, with the goal of creating sustained engagement in the programs.

Koch sees the desired end point of climate change education as addressing the issue of sustainability, which requires that people understand the magnitude of the problem and also change behavior, but she cautioned that most people may never understand the depth of the science. She also sees the need to go beyond the physical, natural, and social sciences to reach people in order to make progress on these issues.

FINAL THOUGHTS

James Mahoney provided closing comments and some thoughts for the future. He pointed to two ideas that came into focus over the two-day discussions: content and values. There was a lot of discussion of the quantity and quality of information at different levels of education. His experience working in the federal government on issues related to acid rain, which, he noted, were similar in some cases to those associated with climate change education, showed him that people working in different areas of research often did not communicate with those outside their area of expertise. He added that this resulted in a poor foundation to "carry the problem through, end-to-end."

Mahoney pointed out that for those teaching climate science and climate change, there is already a large body of work available that, by its nature, is even-handed and transparent and is not focused on advocacy. These resources are designed to give teachers the context in which to teach climate issues, address uncertainties, identify good information, and set appropriate frameworks. He stressed that although this information is not the "last word," it is a very useful resource.

Mahoney closed with a consideration of how society values science: "Do we value [science] as something which really is aimed to give us the best possible answers, albeit uncertain? Do we value science as simply a debating tool?" The issue, from his perspective, is to bring along students, from middle school to college, to an understanding of science as a tool that allows them to better understand earth systems. The concept of uncertainty is at the center of the discussion, he added, but healthy skepticism is not a reason for dismissing science. He emphasized that the goal is to help students understand that scientists strive to get the science and measurements right to the best of their ability, not because they expect to establish the final word on a subject but as part of a process of expanding understanding and reducing uncertainty.

Martin Storksdieck thanked the participants for the rich discussion at the workshop. He emphasized that the goal of the workshop was not to come to conclusions, but to ask and explore complex questions that do not have absolute or clear and easy solutions. Climate change and climate change education raise many complex questions, he added, and the presenters and participants generated a wealth of ideas and possible answers that will be useful in a continued discussion on how to best address the issue of climate change in formal education, K-14.

References

Borick, C.P., Lachapelle, E., and Rabe, B. (2011). *Climate compared: Public opinion on climate change in the United States and Canada*. Governance Studies Program at Brookings. Washington, DC: Brookings Institution.

Boyes, E., Stanisstreet, M., Skamp, K., Rodriguez, M., et al. (2011). *Student mental models of global warming and climate change: What does research tell us?* Presentation at the Workshop on Climate Change Education in Formal Settings, K-14. Climate Change Education Roundtable, Board on Science Education, Division of Behavioral and Social Sciences and Education, Washington, DC.

Buhr, S. (2011). *Navigating climate change in the classroom: Teacher preparation, practices and perceptions*. Presentation at the Workshop on Climate Change Education in Formal Settings, K-14. Climate Change Education Roundtable, Board on Science Education, Division of Behavioral and Social Sciences and Education, Washington, DC.

Chu, H., Shin, D., and Lee, M. (2005). Korean students' environmental literacy and variables affecting environmental literacy. In S. Wooltorton and D. Marinova (Eds.), *Sharing wisdom for our future. Environmental education in action: Proceedings of the 2006 Conference of the Australian Association of Environmental Education* (Chapter 33). Bellingen: Australian Association for Environmental Education.

Erdogan, M. (2009). *Fifth grade students' environmental literacy and the factors affecting students' environmentally responsible behaviors*. Unpublished doctoral dissertation, Middle East Technical University, Turkey.

Erdogan, M., and Ok, A. (2011). An assessment of Turkish young pupils' environmental literacy: Nationwide survey. *International Journal of Science Education, 33*(17), 2,375-2,406.

Hirabyashi, J. (2011). *CLEAN pathways informant survey #1*. Inverness, CA: Inverness Research Associates. Available: http://www.corestandards.org/ [January 2012].

Johnson, R.M. (2011). *National Earth Science Teachers Association survey of the needs and concerns of K-12 earth and space science educators*. Executive summary available: http://www.nestanet.org/cms/content/welcome [June 2012].

King, M.L., Jr. (1948). The purpose of education. *Morehouse College Campus Newspaper*. Available: http://www.stanford.edu/group/King/liberation_curriculum/pdfs/purposeofeducation.pdf [June 2012].

Lappe, M., (2011). *Alliance for Climate Education.* Presentation at the Workshop on Climate Change Education in Formal Setting, K-14. Climate Change Education Roundtable, Board on Science Education, Division of Behavioral and Social Sciences and Education, Washington, DC.

Lynds, S. (2009). *Inspiring Climate Education Excellence (ICEE) program needs assessment survey report.* Boulder: CIRES Education and Outreach, University of Colorado.

Lynds, S. (2010). *ICEE summer workshop 2010 evaluation report.* Boulder: CIRES Education and Outreach, University of Colorado.

Marcinkowski, T. (2010). *Effects of selected EE approaches on environmental literacy and academic achievement: A review for the NCLI Coalition.* Presentation at the 39th Annual Conference of the North American Association for Environmental Education, Buffalo, NY.

Marcinkowski, T. (2011). *Climate literacy results.* Presentation at the Workshop on Climate Change Education in Formal Settings, K-14. Climate Change Education Roundtable, Board on Science Education, Division of Behavioral and Social Sciences and Education, Washington, DC.

Marcinkowski, T., Noh, K., Erogan, M., and Sagy, M. (2011). *Glimpses of climate literacy: Climate literacy as assessed partially by a limited set of items from four recent national assessments of environmental literacy.* Paper prepared for the Workshop on Climate Change Education in Formal Settings, K-14. Climate Change Education Roundtable, Board on Science Education, Division of Behavioral and Social Science and Education, Washington, DC.

McBeth, W., and Volk, T. (2010). The National Environmental Literacy Project: A baseline study of middle grades students in the United States. *Journal of Environmental Education, 41*(1), 55-67.

McBeth, W., Hungerford, H., Marcinkowski, T., Volk, T., and Meyers, R. (2008). *National Environmental Literacy Assessment Project: Year 1, National baseline study of middle grades students. Final report.* Report to the U.S. Environmental Protection Agency, National Oceanic and Atmospheric Administration, and North American Association for Environmental Education under Grant #NA06SEC4690009. Available: http://www.oesd.noaa.gov/pubs_reports/Final_NELA%20minus%20MSELS_8-12-08.pdf [June 2012].

NASA Advisory Council. (1988). *Earth system science: A closer view.* Committee on Earth System Sciences. Washington, DC: National Aeronautics Space Administration.

National Oceanic and Atmospheric Administration. (2004). *The ocean literacy: The essential principles of ocean sciences K-12.* Washington, DC. Available: http://oceanliteracy.wp2.coexploration.org/?page_id=47 [January 2012].

National Research Council. (1996). *National science education standards.* National Committee on Science Education Standards and Assessment. Center for Science, Mathematics, and Engineering Education. Washington, DC: National Academy Press.

National Research Council. (2010). Advancing the science of climate change. America's Climate Choices: Panel on Advancing the Science of Climate Change, Board on Atmospheric Sciences and Climate, Division on Earth and Life Studies. Washington, DC: The National Academies Press.

National Research Council. (2011a). *America's climate choices.* Committee on America's Climate Choices. Board on Atmospheric Sciences and Climate, Division on Earth and Life Sciences. Washington, DC: The National Academies Press.

National Research Council. (2011b). *Climate change education goals, audiences, and strategies: A workshop summary.* S. Forrest and M. Feder, Rapporteurs. Steering Committee on Climate Change Education Goals and Objectives. Board on Science Education, Division of Behavioral and Social Sciences and Education. Washington, DC: The National Academies Press.

National Research Council. (2011c). *A framework for K-12 science education: Practices, crosscutting concepts, and core ideas*. Committee on a Conceptual Framework for New K-12 Science Education Standards. Board on Science Education, Division of Behavioral and Social Sciences and Education. Washington, DC: The National Academies Press.

Negev, M., Sagy, G., Garb, Y., Salzberg, A., and Tal, A. (2008). Evaluating the environmental literacy of Israeli elementary and high school students. *Journal of Environmental Education, 39*(2), 3-20.

Office of the Superintendent of Public Instruction for Washington State. (2009a). *Washington state K-12 science learning standards: June.* Available: http://www.k12.wa.us/Science/pubdocs/WAScienceStandards.PDF [January 2012].

Office of the Superintendent of Public Instruction for Washington State. (2009b). *Washington State K-12 integrated environmental & sustainability learning standards: September 2009.* Available: http://www.k12.wa.us/EnvironmentSustainability/pubdocs/ESE Standards.PDF [January 2012].

Office of the Superintendent of Public Instruction for Washington State. (2011). *Washington state environmental and sustainability literacy plan.* Available: http://www.k12.wa.us/EnvironmentSustainability/pubdocs/WAESLPFinalJuly2011.pdf [January 2012].

Partnership for 21st Century Skills. (2011). *Four leading education organizations.* Washington, DC: Author.

Reiser, B. (2011). *Addressing climate change in the NRC framework.* Presentation at the Workshop on Climate Change Education, K-14. Climate Change Education Roundtable, Board on Science Education, Division of Behavioral and Social Sciences and Education, Washington, DC.

Rodriguez, M., Boyes, E., Stanisstreet, M., Skamp, K., Malandrakis G., et al. (2011). Can science education help to reduce global warming? An international study of the links between students' beliefs and their willingness to act. *International Journal of Science in Society, 2,* 89-100.

Shepardson, D.P., Choi, S., Niyogi, D., and Charusombat, U. (2011). Seventh grade students' mental models of the greenhouse effect. *Environmental Education Research, 17*(1), 1-17.

Shin, D., Chu, H., Lee, E., Ko, H., Lee, M., Kang, K., Min, B., and Park, J. (2005). An assessment of Korean students' environmental literacy. *Journal of the Korean Earth Science Society, 26*(4), 358-364.

Tal, A., Garb, Y., Negev, M., Sagy, G., and Salzberg, A. (2008). Evaluating the environmental literacy of Israeli elementary and high school students. *Journal of Environmental Education, 39*(2), 3-20.

U.S. Global Research Program. (2009). *Climate literacy: The essential principles of climate sciences, a guide for individuals and communities.* Available: http://www.globalchange.gov/resources/educators/climate-literacy [January 2012].

Washington State Legislature. (2009). *Mandatory areas of study in the common school.* (WAC # 392-410-115). Available: http://apps.leg.wa.gov/wac/default.aspx?cite=392-410-115 [June 2012].

Wise, S. (2010). Climate change in the classroom: Patterns, motivations and barriers to instruction among Colorado science teachers. *Journal of Geoscience Education, 58*(5), 97-309.

Appendix A

Workshop Agenda and List of Participants

Workshop on Climate Change Education in Formal Settings, K-14
August 31–September 1, 2011

The workshop will discuss climate change and climate science education in formal settings from Kindergarten through the first two years of college with the goal of building toward innovative practices based on a solid understanding of current trends. The workshop will begin with an investigation on student understanding of climate change and global warming and the state and quality of curricular materials for climate change and climate science in K-14. The broader context for climate change and climate science education will be explored through new generation national and state science standards and the current state of teacher understanding of, and preparation for climate change and climate science education. The workshop will end by featuring and discussing innovative approaches to climate change and climate science education that span into early college.

Wednesday, August 31, 2011

8:15–8:45 Individual Discussions with panelists and commissioned authors
(*Breakfast available*)

8:45–9:00 **Welcome**
Martin Storksdieck (Director, Board on Science Education)
Jim Mahoney (Climate Change Education Roundtable Chair)

9:00–10:15 SESSION 1: INTRODUCTION AND KEYNOTE REMARKS

9:00–9:15 Introduction: Goals for the Workshop
Charles W. "Andy" Anderson (Workshop Committee Chair)

9:15–9:45 Challenges and Opportunities in Climate Change Education
Danny Edelson (National Geographic Society)

Formal education has an important role in preparing citizens to respond appropriately to the challenges posed by climate change. The keynote speaker will address this role and provide an overview of the need for climate change education in schools, the goals for climate education in K-14, and challenges and opportunities inherent to teaching and learning climate change education in schools.

9:45–10:15 Questions and Answers

10:15–10:30 BREAK

10:30–12:00 SESSION 2: STUDENT UNDERSTANDING OF CLIMATE CHANGE

Moderator: *Andy Anderson (Workshop Committee Chair)*

Climate change education is being taught in formal settings in various ways, both within formal courses and other activities within schools (e.g., after-school programs). This session will explore how students currently understand and learn about climate science and climate change, how climate change education is represented in current curricula materials, and appropriate pedagogies that address various goals for climate change education in K-12.

Guiding Questions:

- What does mental model research and select items from the National Assessment of Environmental Literacy suggest about student climate literacy and understanding?
- What is the nature and quality of current materials for teaching climate change and climate science in K-12?
- What are effective teaching strategies for various climate literacy goals?

10:30–11:30 Presentations and Panel Discussions
Eddie Boyes (University of Liverpool): Student Mental Models of Global Warming and Climate Change

Frank Niepold (NOAA): Nature and Quality of Teaching Materials for Climate Change Education

Tom Marcinkowski (Florida Institute of Technology): Climate Literacy and Climate Pedagogy

11:30–12:00 Audience Q&A

12:00–1:00 Continued Audience Discussions
Lunch served

1:00–4:45 SESSION 3: STANDARDS AND TEACHERS

This session will explore two critical aspects that influence the nature and quality of climate change education throughout the K-12 system: how standards may influence what is taught in classrooms, how teachers currently address climate change and climate science, and how teachers can be supported in effective ways.

1:00–2:15 Session 3A: Role of Science Education Standards

Moderator: *Jim Geringer (Workshop Committee Member)*

This section will discuss the role of new science education standards and other frameworks, such as state environmental literacy plans and state standards in providing opportunities for addressing climate change and climate science in the K-12 curriculum.

1:00–2:00 Presentations and Panel Discussions
Brian Reiser (Northwestern University) and *Stephen Pruitt (Achieve)*: Addressing climate change in the NRC *Framework* and the next generation science education standards

Gilda Wheeler (Office of Superintendant of Public Instruction, State of Washington): A perspective from the state of Washington

Stephen Pruitt (Achieve): Challenges with controversial science issues

2:00–2:15 Clarifying Questions to Prepare for the Breakout Discussions

2:15–3:15 Session 3B: Teacher Understanding and Preparation

Moderator: *Tamara Ledley (Workshop Committee Member)*

Teacher preparation and understanding of climate science and climate change issues are key components for providing effective climate change education in K-14. This session will explore current teacher practices in K-14 climate change and climate science education, and strategies to support climate science and climate change teaching in the classroom.

2:15–3:15 Presentations and Panel Discussions

Susan Buhr (University of Colorado at Boulder): Navigating climate science in the classroom: Teacher preparation, practices, perceptions and professional development

Roberta Johnson (National Earth Science Teachers Association): Addressing teacher practices and barriers and challenges inherent with teaching climate change education

Francis Eberle (National Science Teachers Association): Discussant

3:15–3:30 BREAK

3:30–4:30 Breakout Sessions: Small Group Discussions

Workshop participants will continue the discussion initiated in the two previous panel discussions (standards and teacher preparation) during small group discussions. Workshop participants can choose to focus on either the role of standards in climate science and climate change education, or on how teachers are prepared and supported in teaching climate science and climate change.

Guiding Questions:

Role of Science Education Standards

1. What is the role of new Science Education Standards and other frameworks (State Environmental Literacy Plans and State Standards) in providing opportunities or barriers for K-12 CCE? How is the framework similar to or different from current practices?
2. In addition to the areas identified in the Conceptual Framework for New Science Education Standards, where should climate change education be covered in the curriculum?
3. In the translation from the Framework to the Standards, what are the opportunities to embed climate change literacy more broadly across disciplines?
4. What are the leverage points for incorporating climate change education into each level of education (elementary, middle, high school)?

Teacher Understanding and Preparation

1. What types of pedagogical knowledge is needed to teach climate change or climate science? How can we help teachers to obtain the knowledge they need to teach climate change comprehensively?
2. How can teachers and principals overcome skepticism about climate change and climate change education, e.g., from parents or administrators?
3. What are strategies for finding appropriate curricular materials?
4. How can schools/districts organize themselves so that teachers are motivated to teach climate change?

4:30–5:00 **Report from Breakout Session: Synthesis and Lessons Learned**

5:00 **Wrap-up of Day**

Thursday, September 1

8:30–9:00 Individual Discussion of Day 1
(*Breakfast available*)

9:00–9:15 **Welcome and overview of Day 2**
Andy Anderson (Workshop Committee Chair)

9:15–10:45 **SESSION 4: INNOVATIONS IN PROVIDING OPPORTUNITIES TO ENGAGE IN CLIMATE CHANGE EDUCATION IN HIGH SCHOOL AND COLLEGES**

Moderator: *Louisa Koch (Workshop Committee Member)*

This session will explore innovations in teaching climate change education, including links between high school and the first two years of college. Discussion will focus on issues such as student engagement and motivation, addressing the interdisciplinary nature of climate change and climate science, and strategies for education toward stewardship and citizenship.

Guiding Questions:

- What is the role of AP courses, particularly AP environmental science, in teaching students about climate change and climate science?
- What examples of effective and innovative and potentially inter- and transdisciplinary practices in climate change and climate science education can we find in high school and colleges?
- What can we learn from alternative approaches to climate change education in schools that make use of out-of-school models for teaching and learning?

9:15–10:45 **Presentations and Panel Discussions**
Karen Lionberger (College Board–AP Program): AP courses and climate science and climate change education

LuAnne Thompson (University of Washington): Partnerships between high schools and universities

Nicky Phear (University of Montana): Developing and implementing an interdisciplinary climate change minor

Matt Lappe (Alliance for Climate Education): Bringing climate change to schools and back home

Mike Town (Steering Committee Member): Discussant

10:45–11:00 BREAK

11:00–12:00 **Breakout Sessions**
Workshop participants will continue the discussion initiated in the previous panel in small groups, inspired by topics like innovation in high schools, linkages between high school and college, inter- and transdisciplinary approaches, and using out-of-school resources for school-based instruction. The breakout discussions allow participants to innovate and share, but all are asked to address how new ideas can be evaluated and brought to scale.

Guiding Questions: Use questions for overall session (listed above)

12:00–1:00 **Continued Audience Discussions**
Lunch served

1:00–2:00 **Bringing It All Together: A Plenary Discussion**
Moderator: *Andy Anderson (Workshop Committee Chair)*

2:00–2:30 **Workshop Implications and Next Steps**
Andy Anderson (Workshop Committee Chair)
Martin Storksdieck (Director, Board on Science Education)
Jim Mahoney (Climate Change Education Roundtable Chair)

2:30 **Meeting Adjourned**

PARTICIPANT LIST

Workshop on Climate Change Education in Formal Settings, K-14

Bethany Adamec, American Geophysical Union
Charles W. "Andy" Anderson, Michigan State University
John Baek, National Oceanic and Atmospheric Administration
Neela Banerjee, Los Angeles Times/Tribune Co.
Alix Beatty, National Research Council
Miriam Bertram, University of Washington
Jacob Clark Blickenstaff, American Physical Society
David Blockstein, National Council for Science and the Environment
Gillian Bowsen, Monash University
Eddie Boyes, University of Liverpool
Carol Brewer, University of Montana
James Brey, American Meteorological Society
Susan Buhr, Cooperative Institute for Research in Environmental Sciences
Elizabeth Burck, National Aeronautics and Space Administration
David Campbell, National Science Foundation
Carly Carroll, U.S. Environmental Protection Agency
Lin Chambers, National Aeronautics and Space Administration
Nancy Colleton, Institute for Global Environmental Strategies
Juliet Crowell, National Science Resources Center, Smithsonian Institution
Alphonse DeSena, National Science Foundation
Brian Dozd, U.S. Environmental Protection Agency
William Easterling, Pennsylvania State University
Francis Eberle, National Science Teachers Association
Daniel Edelson, National Geographic Society
Akiko Elders, National Science Foundation
Lynn Elfner, The Ohio Academy of Science
Thomas Emrick, Smithsonian Institute
Evelina Feliate-Maurice, U.S. National Aeronautics and Space Administration
Mary Ford, National Geographic Education Programs
Sherrie Forrest, National Research Council
Edward Geary, The Globe Program
Laurie Geller, National Research Council
James Geringer, Environmental Systems Research Institute
Sophia Gershman, Watchung Hills Regional High School
Patricia Gober, Arizona State University
Sara Harris, University of British Columbia
Alexis Heath, National Council for Science and the Environment

Joseph Heimlich, Ohio State University
Matthew Inman, Department of Energy
Roberta Johnson, National Earth Science Teachers Association
Jill Karsten, National Science Foundation
Louisa Koch, Office of Education, National Oceanic and Atmospheric Administration
Jay Labov, National Research Council
Carol Landis, Byrd Polar Research Center
Matt Lappe, Alliance for Climate Education
Tamara Ledley, TERC
Kimberly Lightle, Ohio State University
Karen Lionberger, The College Board
James Mahoney, Consultant
Thomas Marcinkowski, Florida Institute of Technology
Ann Martin, Langley Research Center, National Aeronautics and Space Administration
Erin McDougal, National Science Foundation
Katie McGaughey, National Science Foundation
Cathy Middlecamp, University of Wisconsin–Madison
Kristina Mitchell, Pennsylvania State University
Michael Mogil, How the Weatherworks
Teresa Mourad, Ecological Society of America
Bree Murphy, Estuarine Reserves Division, National Oceanic and Atmospheric Administration
Frank Niepold, National Oceanic and Atmospheric Administration
David Oberbillig, U.S. Department of Energy
Rajul Pandya, University Corporation of Atmospheric Research
Jean Pennycook, National Science Foundation
Nicky Phear, University of Montana
Matthew Pines, National Science Foundation
Monica Plisch, American Physical Society
Stephen Pruitt, Achieve, Inc.
Brian Reiser, Northwestern University
Kimberly Roe, National Science Foundation
Joshua Rosenau, National Center for Science Education
Stacey Rudolph, Office of Education, National Oceanic and Atmospheric Administration
Joel Scheraga, U.S. Environmental Protection Agency
Karen Scott, U.S. Environmental Protection Agency
Bono Sen, *Environmental Health Perspectives*, National Institute of Environmental Health Sciences
Jennifer Skene, Lawrence Hall of Science, University of California, Berkeley

Nancy Songer, University of Michigan
Peg Steffen, National Oceanic and Atmospheric Administration
Martin Storksdieck, National Research Council
Daniel Strain, *Science News*
Cathlyn Stylinski, University of Maryland Center for Environmental Science
Marilyn Suiter, National Science Foundation
Surili Sutaria, Association of State and Territorial Health Officials
Luanne Thompson, University of Washington
Mike Town, Redmond High School
Jeanne Troy, Koshland Science Museum
Jermelina Tupas, National Institute of Food and Agriculture, U.S. Department of Agriculture
Louie Tupas, National Institute of Food and Agriculture, U.S. Department of Agriculture
Elizabeth Walsh, University of Washington
Cynthia Wei, National Science Foundation
Ming-Ying Wei, National Aeronautics and Space Administration
Gilda Wheeler, Washington State Office of the Superintendent of Public Instruction
Ted Willard, American Association for the Advancement of Science
Carolyn Wilson, National Science Foundation
Linda Wilson, Project 2061/AAAS
Deborah Wojcik, University of Florida

Appendix B

Climate Change Education Roundtable

James Mahoney (*Chair*), Environmental Adviser
Charles W. "Andy" Anderson, Michigan State University
David Blockstein, National Council for Science and the Environment
F. Stuart Chapin III, University of Alaska
Caron Chess, Rutgers University
William Easterling, Pennsylvania State University
Lynn Elfner, Ohio Academy of Science
James E. Geringer, Environmental Systems Research Institute, Inc.
Patricia Gober, Arizona State University
Joseph Heimlich, Ohio State University
Roberta Johnson, National Earth Science Teachers Association
Tamara Ledley, Center for Science Teaching and Learning, TERC, Inc.
Anthony Leiserowitz, Yale University
Robert Lempert, RAND
Michael McElroy, Harvard University
Janet Peace, Pew Center on Global Climate Change
Walter Staveloz, Association of Science and Technology Centers
Will Travis, San Francisco Bay Conservation and Development Commission

Ex Officio

David Campbell, National Science Foundation
Gregory Crosby, U.S. Department of Agriculture

Jill Karsten, National Science Foundation
Louisa Koch, National Oceanic and Atmospheric Administration
Michael Lach, U.S. Department of Education
Robert O'Connor, National Science Foundation
Joel Scheraga, U.S. Environmental Protection Agency
Bill Valdez, U.S. Department of Energy
Ming-Ying Wei, National Aeronautics and Space Administration

NRC Staff

Michael A. Feder, *Study Director* (until February 2011)
Sherrie Forrest, *Study Director* (since March 2011)
Martin Storksdieck, *Director, Board on Science Education*
Paul G. Stern, *Senior Scholar, Board on Environmental Change and Society*
Rebecca Krone, *Program Associate*
Anthony Brown, *Senior Program Assistant*

Appendix C

Biographical Sketches of Presenters, Steering Committee Members, and Staff

PRESENTERS

EDDIE BOYES is senior lecturer in education and chairman of the Board of Studies of the Centre for Lifelong Learning and director of postgraduate research in the Educational Development Division at the University of Liverpool. He is involved with a number of projects in the division, mainly involving statistical analysis. His current research interests include conceptual understanding of physical phenomena and the preconceptions that children, students, and adults hold about major environmental and health issues, including public concerns about scientific advances. He is a member of the Environmental Education Research Unit and has published widely on children's understanding of science and environmental education issues.

SUSAN BUHR directs the education outreach program of the Cooperative Institute for Research in Environmental Sciences (CIRES). Before beginning her work in K-12 education, she conducted research in atmospheric chemistry analytical methods with CIRES and the Aeronomy Laboratory of the National Oceanic and Atmospheric Administration. She enjoys the opportunity to learn about a wide variety of science topics through education work, as well as working with educators, students, geoscientists, and social scientists. Her current projects include professional development workshops for science teachers, provision of education related to research projects, and oversight of numerous other education projects within the CIRES outreach group. She has a B.S. in chemistry from California Polytechnic State University, San Luis Obispo, and a Ph.D. in analytical chemistry from the University of Colorado, Boulder.

FRANCIS Q. EBERLE is executive director of the National Science Teachers Association. Previously, as executive director of the Maine Mathematics and Science Alliance (MMSA), he worked to develop state curriculum frameworks and provide professional development and resources to schools and teachers throughout Maine. Prior to joining MMSA, he was an adjunct faculty member of the University of Southern Maine and is also a former Maine middle and high school science teacher. He was president of the Maine Science Teachers Association and has served on advisory groups for the National Alliance of State Science and Mathematics Coalitions, the Centers for Ocean Sciences Education Excellence, and the Maine Space Grant Consortium. His research has focused on integrating engineering into the high school curriculum, training in-service teachers, mentoring new teachers, involving parents in science and mathematics, and integrating technology into the science and mathematics classroom. He has a B.A. in science education from Boston University, a master's degree in educational psychology from the University of Connecticut, and a Ph.D. in educational studies from Lesley University.

DANIEL EDELSON is vice president for education at the National Geographic Society and executive director of the society's Education Foundation. In his position as vice president, he oversees National Geographic's outreach to educators and its efforts to improve geographic and geoscience education in the United States and abroad. Previously, he was a professor in education and computer science at Northwestern University. He also created professional development programs for educators from middle school through college and led several large-scale instructional reform efforts in the Chicago Public Schools. He has written extensively on motivation, classroom teaching and learning, educational technology, and teacher professional development. He is author of numerous papers in journals, edited books, and conference proceedings, including *The Cambridge Handbook of the Learning Sciences*, and *The International Handbook on Science Education*, among others. He has a B.S. in engineering sciences from Yale University and a Ph.D. in computer science (artificial intelligence) from Northwestern University.

ROBERTA JOHNSON is the executive director of the National Earth Science Teachers Association (NESTA) and director of Special Projects at the University Corporation for Atmospheric Research Office of Education and Outreach. She is also a research scientist in the High Altitude Observatory at the National Center for Atmospheric Research. NESTA is a nonprofit educational organization that works to advance and improve earth science education at all levels. Previously, as a research scientist at the University of Michigan, she started *Windows to the Universe*, an award winning web-based educational tool. She serves on numerous advisory boards for projects in science education, outreach, and diversity and has extensive experience advising the National Aeronautics and Space Administration, the National Science Foundation, and a variety of professional societies. She is the chair of the International Council for Science's Ad Hoc Review Panel on Science Education. She has B.S., M.S., and Ph.D. degrees, the latter in geophysics and space physics, from the University of California, Los Angeles.

MATT LAPPE is a program officer at the Alliance for Climate Education (ACE), Colorado. Before joining ACE, Mr. Lappe worked as a policy analyst for the Tomales Bay Institute, where he helped Peter Barnes develop the Cap and Dividend climate policy framework, now advocated by politicians across the country. He taught at a small charter high school in Mendocino County, where he headed the science and social studies departments, and he founded the Sustainable Energy Education Program. He has a B.S. and an M.S. from Stanford University's Earth Systems Program; he has also studied paleoclimate and environmental hydrology throughout Patagonia, Vietnam, and Cambodia.

KAREN LIONBERGER is the director of curriculum and content development for the College Board's Advanced Placement (AP) environmental science and AP physics courses. She taught AP environmental science for many years in Atlanta, where she also worked with the Georgia Department of Education as a workshop consultant for AP teachers. For six years, she served as an AP environmental science exam reader for the College Board. She has worked on numerous projects as a coauthor and content editor for instructor's guides and student study guides that support the AP environmental science curriculum. In 2010, she coauthored a new student study guide called *Fast Track to a Five: Preparing for the AP Environmental Science Exam.*

THOMAS MARCINKOWSKI is the Acopian program chair of the graduate program in environmental education at the Florida Institute of Technology and coordinator of the university-wide undergraduate Quality Enhancement Program. He is active in the efforts of the North American Association for Environmental Education (NAAEE) to enhance the preparation and professional development of environmental education, serving on the writing team and as a reviewer for the National Council for the Accreditation of Teacher Education Standards, as secretary to the certification advisory council, and as chair of the accreditation board. He also served as chair of NAAEE's research commission and contributed reviews of research. He has been a member of the National Environmental Literacy Assessment (NELA) research team. He is currently developing a framework for assessing environmental literacy with representatives from NELA, NAAEE, and the Programme for International Student Assessment. He has an M.S. in forestry, with a concentration in nonformal environmental education and environmental interpretation, and a Ph.D. in curriculum and instruction, with a concentration in science and environmental education, from Southern Illinois University.

FRANK NIEPOLD is climate education coordinator in the Climate Program Office of the National Oceanic and Atmospheric Administration (NOAA); a member of the NOAA Education Council; cochair of the newly formed Education Interagency Working Group of the Climate Change Science Program (CCSP); member of the Communications Interagency Working Group; and a founding member of the Climate Literacy Network. At NOAA, he develops and implements climate goal education and other efforts that specifically relate to NOAA's environmental literacy crosscutting priority. He is coauthor of *Climate Literacy: The Essential Principles of Climate Science*. As cochair of the CCSP's education interagency working group, he works to develop the interagency partnership as well as coordination and strategic direction of the federal climate science education efforts to support the development of a knowledgeable and informed nation relative to climate. He has a B.A. in human ecology from the College of the Atlantic in Bar Harbor and an M.S.Ed. in earth space science education from Johns Hopkins University, with areas of concentration in earth observing systems, scientist/teacher/student collaboration, and earth systems science education focused on climate.

NICKY PHEAR is a faculty member at the University of Montana, where she coordinates and instructs in the Climate Change Studies Program. The climate change studies minor offers students a multidisciplinary understanding of climate change and involves them in developing solutions. She coteaches the introductory course Climate Change: Science and Society and develops experiential learning opportunities for students through internships, practicums, and field courses. She has taught for several campus- and field-based programs, including the university's Wilderness and Civilization Program, the Wild Rockies Field Institute, Prescott College, and the Colorado Outward Bound School. She is the cofounder of a summer "Cycle the Rockies" program, in which university students study alternative energy production and climate change as they bicycle 700 miles across the state of Montana. In the winter, she leads a field course in Vietnam's Mekong Delta about climate change impacts and adaptation. She has an M.S. in environmental studies from the University of Montana and is pursuing a Ph.D. in sustainability education through Prescott College.

STEPHEN PRUITT is the vice president of content, research, and development at Achieve, Inc., which he joined as director of science in July 2010. He continues to lead the development of the *Next Generation Science Standards*. He began his career as a high school chemistry teacher in Georgia, where he taught for 12 years. In 2003, he joined the Georgia Department of Education as the program manager for science, becoming director of academic standards four years later, overseeing the continued implementation of the Georgia Performance Standards in all content areas. In 2008, he became the associate superintendent of assessment and accountability, responsible for directing all state assessments and overseeing the No Child Left Behind accountability process. In April 2009, he became chief of staff to the state school superintendent, coordinating the work of the agency and a variety of projects, including Georgia's third-ranked Race to the Top. He also served as president of the Council of State Science Supervisors and a member of the writing team for the College Board's *Standards for College Success* science standards. He has a B.S. in chemistry from North Georgia College and State University, an M.Ed. in science education from the University of West Georgia, and a Ph.D. in chemistry education from Auburn University.

BRIAN REISER is professor of learning sciences in the School of Education and Social Policy at Northwestern University. His research examines how to make scientific practices, such as argumentation, explanation, and modeling, meaningful and effective for classroom teachers and students. He leads the MoDeLS project (Modeling Designs for Learning Science), which is developing an empirically based learning progression for the practice of scientific modeling, and BGuILE (Biology Guided Inquiry Learning Environments), which is developing software tools for supporting students in analyzing biological data and constructing explanations. He is also on the leadership team for IQWST (Investigating and Questioning our World through Science and Technology), a collaboration with the University of Michigan that is developing a middle school project-based science curriculum. He was a founding member of the first graduate program in learning sciences created at Northwestern and chaired the program from 1993 to 2001. He was co-principal investigator in the Center for Curriculum Materials in Science, exploring the design and enactment of science curriculum materials. He also served on the editorial boards of

Science Education and the *Journal of the Learning Sciences*. He has a Ph.D. in cognitive science from Yale University.

LUANNE THOMPSON is director of the University of Washington Program on Climate Change as well as professor of oceanography and adjunct professor of physics and atmospheric sciences. Her research program focuses on the ocean's role in climate, using ocean and climate models along with satellite data. She is a senior fellow in the Joint Institute for the Study of Atmosphere and Oceans and a global health and environment fellow. As director of the University of Washington's Program on Climate Change, she leads the graduate certificate in climate sciences and the undergraduate minor in climate sciences. She has taught classes at both the graduate and undergraduate levels on ocean physics, the ocean's role in climate, climate dynamics, and climate modeling. She has a B.S. in physics from the University of California, Davis, an M.A. in physics from Harvard University, and a Ph.D. from the Massachusetts Institute of Technology joint program in oceanography and oceanographic engineering.

GILDA WHEELER is the program supervisor for environmental and sustainability education in the Washington State Office of the Superintendent of Public Instruction (OSPI). She is responsible for supporting districts, schools, teachers, and students in implementing legislatively mandated environmental and sustainability education in Washington state. She also serves on a number of state and national boards and committees, including as cochair of the E3 Washington K-12/Teacher Education Sector steering committee, the national K-12 sector of the U.S. Partnership for Education for Sustainable Development, and the Council of Chief State School Officers' EdSteps Global Competency work group. Prior to joining OSPI, she was the program director for the nonprofit education organization Facing the Future, where she developed hands-on experiential curricula on global sustainability issues and led teacher workshops around the country. She was a classroom teacher for many years before turning to efforts that would support all teachers and advance the field of environmental and sustainability education. She has a B.A. in geography and an M.Ed. from the University of California, Santa Barbara.

STEERING COMMITTEE AND STAFF

CHARLES W. "ANDY" ANDERSON is professor in the Department of Teacher Education at Michigan State University. His current research focuses on the development of learning progressions leading to environmental science literacy for K-12 and college students. He has used conceptual change and sociocultural research on student learning to improve classroom science teaching and science teacher education, science curriculum, and science assessment. He is past president of the National Association for Research in Science Teaching and has been coeditor of the *Journal of Research in Science Teaching* and associate editor of *Cognition and Instruction*. He served as a member of the Science Framework Planning Committee of the National Assessment of Educational Progress (NAEP) and the NAEP Science Standing Committee. He has a B.A. in chemistry, an M.A. in science education, and a Ph.D. in science education from the University of Texas at Austin.

CAROL BREWER is professor emeritus of biology at the University of Montana and associate dean of the College of Arts and Sciences. Her research program focused on physiological plant ecology and ecological education. She founded the consulting group Prairie Ecotone Research Group, LLC. She served on the editorial boards of *Conservation Biology* and *Frontiers in Ecology and the Environment*. She was the vice president of education and human resources of the Ecological Society of America from 2000 to 2006. She led education planning for the National Ecological Observatory Network and the National Phenology Network, and she currently serves on boards of the American Institute of Biological Sciences, the Longterm Ecological Research Network (as chair), Earth and Sky Radio, and the National Ecological Observatory Network. In 2007, she received both the Eugene P. Odum Award for Ecological Education from the Ecological Society of America and the Education Award from the American Institute of Biological Sciences. She has a B.A. in biology from California State University, Fullerton, as well as a B.S. in science education, an M.S. in zoology and physiology, and a Ph.D. in botany from the University of Wyoming.

LYNN ELFNER is chief executive officer at the Ohio Academy of Science in Columbus. He has also worked at the Mt. Orab Local School District, Ohio State University, the Ohio Environmental Council, and the Ohio Office of Budget and Management. He is a fellow of the American Association for the Advancement of Science and has received many awards, including the Honorary 100 from Ohio in Natural Resources; the Centennial honoree and Friend of Science Award from the Science Education Council of Ohio; the President's Award from the Ohio Alliance for the Environment; and the President's Award from Ohio School Boards Association. Current activities include archivist and board of directors of the National Association of Academies of Science; councilor of the Ohio State Chapter of Sigma Xi; ex officio member of the board of trustees of the Ohio Academy of Science; and member of the board of directors, ex officio alternate member, and board of trustees of the Ohio Historical Society. He has a B.S. and an M.S. in botany from the Ohio State University.

SHERRIE FORREST (*Study Director*) is an associate program officer with the Ocean Studies Board and the Board on Science Education at the National Research Council. She currently provides support on several projects, including the Roundtable on Climate Change Education, the Conceptual Framework for New Science Education Standards, and the Effects of the Deepwater Horizon Mississippi Canyon-252 Oil Spill on Ecosystem Services in the Gulf of Mexico. Previously, she worked as a freelance science writer. Before transitioning to her current path, she worked in the development and production of feature films and documentaries in California and New York. She has a B.A. in English literature from Pepperdine University and an M.S. in biological oceanography from the Institute of Marine and Coastal Sciences at Rutgers University.

JAMES E. GERINGER, the 30th governor of Wyoming, is the director of policy and public sector strategies for the Environmental Systems Research Institute (ESRI) in Redlands, California. From 1967 to 1977, he served in the U.S. Air Force. He has also worked at the Missouri Basin Power Project's Laramie River Station. In 1982, he successfully ran as a Republican for a seat in the Wyoming House of Representatives. After serving there for six years, he won a seat in the Wyoming Senate. In 1994, State Senator Geringer was elected as Wyoming's governor. As governor, he helped pass laws that regulated class action lawsuits, reformed bankruptcy laws, toughened crime laws, legalized charter schools, and lowered taxes. However, he broke with the Republican Party in supporting environmental rulings and the Equal Rights Amendment. He is one of the founding governors of Western Governors University and is currently chairman of its board of trustees. He has a B.S. in mechanical engineering from Kansas State University.

LOUISA KOCH is director of the Office of Education of the National Oceanic and Atmospheric Administration (NOAA), which is responsible for educating the public about the role of the ocean, the coasts, the Great Lakes, and the atmosphere in the global environment and developing the next generation of professionals capable of understanding and managing those resources. As director, she chairs the Education Council, which consists of education directors from all major education programs in NOAA. She served as NOAA's deputy assistant administrator for research. Before joining NOAA, she served as the commerce branch chief at the U.S. Office of Management and Budget and was a presidential management intern at the Department of Defense and an economist with the Joint Economic Committee of the U.S. Congress. She has a B.S. in physics from Middlebury College and an M.S. in electrical engineering from the Massachusetts Institute of Technology.

TAMARA SHAPIRO LEDLEY is senior scientist and interim director of the Center for Science Teaching and Learning at the Technical Education Research Center. Her work in earth systems science education included developing museum exhibits, contributing science content to planetarium shows, directing teacher training programs, developing learning activities and a teacher guide for the Global Learning and Observations to Benefit the Environment (GLOBE) Program, and facilitating the use of earth science data in educational contexts. She developed the Earth Exploration Toolbook; led the Digital Library for Earth System Education Data Services and AccessData projects; and led the Tools for Data Analysis in the Middle School Classroom project. She is a founding member and chair of the Climate Literacy Network, with projects that include the Climate Literacy and Energy Awareness Pathway. She has served as chair of the Standing Committee for Education and as vice president for the Federation of Earth Science Information Partners (ESIP Federation) and is a member of the board of trustees of the Foundation of Earth Science. She was chair of the Committee on Global and Environmental Change of the American Geophysical Union. She has a B.S. in astronomy from the University of Maryland and a Ph.D. in meteorology and physical oceanography from the Massachusetts Institute of Technology.

MARTIN STORKSDIECK is the director of the Board on Science Education and of the Roundtable on Climate Change Education at the National Research Council. Previously, he was director of project development and senior researcher at the Institute for Learning Innovation (ILI) where he directed ongoing research studies of science learning in immersive environments; models of involving researchers and scientists in science museums and science centers; and the impact of science hobbyists, such as amateur astronomers, on the public understanding of science. Prior to that, he was a science educator with a planetarium in Germany, where he developed shows and programs on global environmental change; served as editor, host, and producer for a weekly environmental news broadcast; and worked as an environmental consultant specializing in local environmental management systems. He has an M.S. in biology from the Albert-Ludwigs University in Freiburg, Germany, an M.P.A. from Harvard University's John F. Kennedy School of Government, and a Ph.D. in education from Leuphana University in Lüneburg, Germany.

MICHAEL TOWN is an Einstein fellow with the National Science Board of the National Science Foundation and an Advanced Placement environmental science teacher at Redmond High School in Redmond, Washington. His research specializes in fire ecology and the ecological interaction between pine beetles and lodgepole pines in the Yellowstone area. He has written significant environmental and science, technology, engineering, and mathematics curricula. The most notable curriculum is the Cool School Challenge, which enables students to conduct energy audits in schools across the United States. He has been recognized with numerous awards, including the National Education Association Foundation's Green Prize for the United States; Environmental Educator of the Year from the North American Association of Environmental Educators; the Pemco/KCTS Golden Apple Award; the Conservation Fund Environmental Educator Award for the United States; the Western Washington/Huxley College Distinguished Alumni Award; the Amgen Science Teacher Award; the AP/Siemens Math/Science Teacher of the Year for Washington State; and the Cox/KIRO TV Environmental Hero award. He has a B.S. in science education from Western Washington University and an M.A. in education from the University of Washington.